Beauty and the Bees

By

Dr Sara Robb

for Viv,

with my best wishes.

Dr Sara Robb

MEG

Northern Bee Books

ISBN 978-1-908904-13-3

Northern Bee Books

Published 2012
Ruxbury Publications

Set in Gill Sans

Design and Artwork
D&P Design and Print
Worcestershire

Beauty and the Bees

By

Dr Sara Robb

Measurement Equivalents
Units of Volume
I tsp – teaspoon = 5 milliliters
I T- tablespoon = I5 milliliters
¼ C – ¼ cup = 60 milliliters
½ C – ½ cup = I20 milliliters
I C- cup = 240 milliliters
I ml- milliliter = I/5 teaspoon
5 ml- 5 milliliters = I teaspoon
I0 ml- I0 milliliters = 2 teaspoons
I5 ml- I5 milliliters = 3 teaspoons or I tablespoon

Units of Mass or Weight
I mg- I milligram = 0.00I gram
I g – I gram = I000 milligrams
I oz- I ounce = 28.35 grams
w/w % - weight to weight percentage = weight of solute/ (weight of solute + weight of solvent) * I00

Beauty and the Bees

By

Dr Sara Robb

Table of contents

List of Tables and Figures

About the Author

Dr Sara Robb is an award-winning scientist, professional speaker, author and managing director of her company Bath Potions. Sara grew up in Iowa and began her education at Iowa State University. Continuing her education, Sara studied pharmacognosy and toxic natural products at the Philadelphia College of Pharmacy and Science, before moving to Hershey Medical School for her PhD. While at Hershey Medical School, Sara focused on aging and the role of oxidative stress in the process. For her PhD research, she won the Marian Kies Memorial Award, from the American Society for Nuerochemistry, for outstanding research conducted during graduate training. Following her time in Hershey, Sara moved to the United Kingdom to continue her research. In 2003, Sara left academic research and founded her company Bath Potions, which specialises in honey soaps and beeswax creams. Recipes are available in her first book, *Dr Sara's Honey Potions*. Sara lives in North London with her family and runs workshops at the British Beekeepers Association Spring Convention and The National Honey Show each year. Sara has a continued interest in aging and anti-oxidants, particularly, the anti-oxidants in bee products and how they can slow the aging process.

Preface

Oxidative stress plays a role in aging and disease, with cellular damage, aging and death delayed by anti-oxidants. Bee products contribute to health and beauty through their strong anti-oxidant activity. Specifically, honey, propolis, and royal jelly contain polyphenol anti-oxidants. The anti-oxidant activity of bee products vary, however all are capable of neutralising hydrogen peroxide, superoxide and hydroxyl radical, the oxidants responsible for causing aging. Anti-oxidant levels in honey can be predicted using honey colour; the darker the honey, the higher the anti-oxidant content. The level of anti-oxidants increase in the following honeys as colour increases: acacia, lime, manuka, heather, honeydew and buckwheat.

Honey is useful as a dietary supplement, because honey increases physiological levels of anti-oxidants. *Beauty and the Bees* contains a collection of sweet and savoury honey recipes. These recipes contain higher levels of anti-oxidants, vitamins, minerals, proteins and amino acids than the same recipe made with sugar alone. Dietary anti-oxidants, such as the polyphenols in honey, can slow the aging process. Substituting some of the sugar in your diet with honey is an easy way to increase your dietary intake of these age defying anti-oxidants. Using dark honey in recipes will further increase the levels of anti-oxidants due to their higher levels of polyphenol anti-oxidants.

As well as containing anti-oxidants, bee products also have anti-bacterial and anti-viral activity, as well as other healing properties. Many home remedies use beeswax, propolis and honey as active ingredients. Aunt Bea's remedies treat colds, sore throats, hay fever and help skin problems, such as eczema and psoriasis. We now have the scientific knowledge to understand their efficacy. Finally, cosmetics containing anti-oxidants can help reduce the signs of aging by decreasing the oxidative damage to our skin. Honey, propolis, and beeswax are good sources of functional anti-oxidants for beauty products. Bee products applied to the skin decrease the signs of aging and protect skin from oxidative stress. Using beauty products made with honey, beeswax and propolis can keep you looking young and beautiful.

I have a long-standing interest in oxidative stress and aging, particularly the chemistry of free radicals. When I began my thesis research in the Department of Neuroscience and Anatomy, at Pennsylvania State University's Hershey Medical School, I decided I would like to focus my studies on how oxidative stress damages cells. Oxidative stress not only plays a role in disease, but controls aging. My research focused on the mechanisms involved in the aging process and the changes that occur in cells as we age. I received my PhD in 1998 for my research examining how oxidative stress damages cells and how anti-oxidants prevent these changes. I enjoyed my research very much and received the Marian Kies Award, for outstanding thesis research, from the Society for Neurochemistry for my work. In 1998, I moved to the United Kingdom to continue my scientific research.

My main interest was still how oxidative stress caused cellular damage and aging.

I enjoyed living in the United Kingdom and became involved with an Englishman. A few years passed, I married my Englishman and we had a baby. Jasmine was born in 2003. I spent my maternity leave learning how to make soap. Baby Jasmine sat in her little chair while I mixed up the soap. When it was time for my maternity leave to end, I decided I would leave my career as a scientist behind in order to spend more time with my baby girl, Jasmine. It was at this point I set up my business, Bath Potions. I continued to make soaps and develop products. The products I made were very nice but there was nothing magical about them. Something was missing. The magical, missing ingredient was honey.

I began making products with honey when Jasmine was almost a year old. A friend gave me a jar of heather honey as a gift. This jar of honey proved to be an inspiration and changed the focus of my business. The first honey soap I made contained a spoonful of the gifted heather honey. The heather honey transformed my soap into something fabulous. I could tell there was something different as I was making it. The soap had a wonderful caramel aroma as it processed. I could not wait to try the soap. Jasmine needed a bath that day; the perfect opportunity to test my new heather honey soap. Jasmine played happily in the bath unaware that she was testing the premiere soap for my new product line. Jasmine enjoyed her bath very much and approved of the new honey soap. When I took Jasmine out of the bath, her skin felt like velvet. I could not believe the addition of a small spoonful of honey could transform my soap from a good product to an exceptional product. Once I started to use honey soap on Jasmine I did not need to use any baby oil or moisturising lotion to keep her skin soft. The honey seemed to have magical properties. I have since been on an exciting journey into the world of bees and honey.

Years have passed since I first made soap with honey. In this time, I have had another daughter, Meggy Jayne, and developed other products including honey soaps, beeswax lip balms and beeswax moisture creams. Honey is the focus of my business, specialising in cosmetics made with honey and beeswax. The more I worked with honey and beeswax, the more interested in bees and honey I became. I began to attend the National Honey Show and the British Beekeeping Association Convention. Through the years, I have met many beekeepers and listened to their stories about bees and honey. I also began to read the scientific research about bees and bee products as anti-oxidants. This rekindled my interest in oxidative stress, aging and the mechanisms by which honey and other bee products can act as anti-oxidants. As I learned more about the health benefits of eating honey and the qualities bee products add to cosmetics, honey became more of my family's life. Toiletries containing honey and other bee products have replaced all of the products previously used in our house. Jasmine and Meggy Jayne use honey soaps and beeswax creams as a part of their everyday routine.

My first book, Dr Sara's Honey Potions, was well received, so I decided to write a second book. In concept, the second book began as a recipe book, which was really to be a continuation of *Dr Sara's Honey Potions*. I was speaking to my mother about my new book when she suggested I add some of our family recipes using honey and beeswax as ingredients. My great-grandfather, George, had been a beekeeper and my family has been cooking with honey and making home remedies with bee products for four generations. My mother persuaded me to include our family recipes. I think I was most nervous about including food recipes. I am a scientist, not a chef! I have done my best to write the food recipes in a way I hope makes sense to those of you who are cooks. Most of the recipes in *Beauty and the Bees* been passed down from generation to generation in our family. The recipes are hearty, home recipes and include sweet and savoury honey recipes, beeswax and honey beauty recipes and Aunt Bea's remedies.

The decision to include our family recipes meant that I needed to find and compile them. Many of the recipes I had to hand, some I had not seen for years, while others were mere memories. I had a few of my grandmother's recipes that I copied years ago, but remembered that my grandmother once had a leather satchel that contained all of the "receipts" she had collected throughout her life. I remember my grandmother using the recipes she kept in this folder when I visited my grandparents when I was a child.

Looking through my grandmother's recipes brought back memories of the summer I spent with my grandparents when I was 12 years old. My family's house in Iowa was hit by a tornado. The damage to our house was extensive and we could not live there while the builders repaired the damage. My grandparents suggested that I come to stay with them in New York for the summer. Going away for the summer was very exciting. I was going on an airplane all by myself and spending my first summer away from my parents. When I arrived, I found Windham to be a beautiful, hilly place, so very different from the flat plains of the Midwest. That summer I had an opportunity to find out all about my grandmother's childhood in New York. I probably asked her too many questions but now I wish I had asked more.

My grandmother, Florence, born in 1899 in Blenheim, lived her whole life in upstate New York. Her father, George, kept bees when she was a child in Blenheim. Florence would help her father with his bees and help her mother cook the families meals. My great-grandmother, Nellie used honey or a combination of honey and sugar in her cooking. They used honey to sweeten tea and coffee, to make cookies, desserts and candies. George kept his bees by the buckwheat fields, near where blueberries grew wild in the hills. The area was full of beautiful wildflowers and proved a wonderful site for beehives. George produced a number of varieties of honey including buckwheat honey, clover honey and wildflower honey. Grandma said there was always a bottle of George's honey on the dining table when she was a little girl and honey was eaten on bread at nearly every meal.

My grandmother continued the tradition of cooking with honey through her life. The summer I spent with my grandparents, my grandma and I made many of Nellie's recipes, including coleslaw, cookies and rhubarb sauce. Most of the recipes my grandmother made without using a recipe however, there were a few grandma did not make very often and she needed to "check the receipt". Grandma's worn, brown-leather folder which contained all her "receipts" was kept on the top shelf in the larder, next to the sugar and flour canisters. Grandma's receipt folder contained countless small pieces of yellowed paper each with a recipe carefully written on it.

My grandmother also told me about the home remedies they made with bee products. My great-grandfather's sister, Beatrice prepared the family's remedies. Nellie always had a medicine bottle with Aunt Bea's oxymel in the pantry. When children developed coughs or had sore throats, they took the brown glass bottle of oxymel down from the medicine shelf, and administered the remedy by the spoonful. Aunt Bea's remedies now have scientific research which validates their efficacy.

Remarkably many of these old recipes and remedies were full of anti-oxidants, vitamins and minerals because of the bee products they contained. I am convinced the use of bee products helped my family live long, healthy lives. My journey into the world of honey has taken me full circle back to my interest in oxidative stress and aging, with a new focus on the anti-oxidants in bee products, including honey.

Beauty and the Bees begins with an Introduction by nutritionist Domingo J. Piñero, discussing the importance of honey as a functional food and how honey, as a part of a nutritious diet, can keep you young and healthy. The aging process is discussed in Chapter 1. The cause of aging is oxidative damage and anti-oxidants, in our diets and in our beauty products, can decrease the signs of aging. Honey, beeswax, propolis, royal jelly and bee bread are all rich in anti-oxidants and can be used to fight oxidative stress. Chapter 2 examines the anti-oxidants in bee products and their value as dietary supplements, functional foods, and in home remedies and cosmetics. The bee product with the broadest usage is honey. Honey is a functional food, a supplement, and an active ingredient in remedies and beauty products. Chapter 3 examines the levels of polyphenol anti-oxidants, in a selection of mono-floral honey. Dietary anti-oxidants can slow the effects of aging. You can use the colour of honey to select honey varieties high in anti-oxidants, the darker the colour, the higher the anti-oxidants. Substituting some of the sugar in your diet with honey will significantly increase your physiological levels of anti-oxidants and can slow the aging process. Recommendations for incorporating honey into culinary recipes are made in Chapter 4.

The recipes in Chapter 5 through Chapter 8 are culinary recipes designed to increase dietary anti-oxidants. Suggestions on how to start your day with honey anti-oxidants are in Chapter 5. While most of us think of sweet food when we think of food containing honey, honey adds flavour and anti-oxidants to savoury dishes as well. Chapter 6 contains recipes for savoury dishes, including tasty starters and delicious main dishes. We all enjoy desserts and sweet treats now and then. Baked goods and confection recipes, made with honey, feature in Chapter 7 and Chapter 8. Baked goods and confections represent an easy opportunity to replace sugar with anti-oxidant rich honey. Adding honey to recipes increases the levels of vitamins, minerals, proteins, amino acids and anti-oxidants in the recipe compared to the same treat made with sugar alone.

As well as increasing anti-oxidants in food, bee products have had a place through history in remedies and other home products. Beeswax, propolis and honey have medicinal qualities that lend themselves to home remedies. Recipes for Aunt Bea's Remedies are in Chapter 9, including remedies for colds, skin conditions and even to help the little ones off to sleep. In addition to use in home remedies, honey, propolis and beeswax have also been very important ingredients in other cosmetics, such as soaps and moisture creams. The spotlight of Chapters 10 and 11 is on how to incorporate honey, beeswax and propolis into soaps and creams for added healing properties and anti-aging activity. The anti-oxidant properties of honey, beeswax and propolis greatly improve toiletries made with these functional ingredients.

The recipes in *Beauty and the Bees* are written in US, Imperial and metric units. Appendix 1 is a handy reference to convert between these units and contains information about volume and weight measurements as well as temperature conversions between Fahrenheit and Celsius. Appendix 2 is taken from Jonathan W. White's *Composition of American Honeys* (1962). A selection of honey varieties are listed by honey colour. While this appendix includes data collected on only US honeys, it is a good source of information about honey colour by botanical source, which can be used as a predictor of the levels of anti-oxidants in honey varieties.

Acknowledgements

First, I would like to acknowledge my family. While the recipes in Beauty & the Bees are family recipes, we had not made many of them for a long time. I appreciate the help my mother gave me in finding copies of the recipes and for searching her memory on many occasions to help me. I must acknowledge Sandip, Jasmine and Meggy Jayne for testing all the recipes for this book. I cannot tell you how much honey we consumed testing the recipes, but I am sure we all increased our anti-oxidant capacities, significantly! My daughter, Meggy Jayne was my assistant in the kitchen and was fantastic help. Jasmine also helped in the kitchen but was usually more interested in tasting than making. I also must thank my husband, Sandip for setting out my bee slippers, making my coffee and feeding me, during the preparation of the manuscript. I would also like to acknowledge Sandip and Chi Li for useful discussion during the research stage.

I am grateful to nutritionist, Dr Domingo Pinero for writing an introduction to this book and his knowledge about functional foods and support for the project. I would also like to thank The National Honey Board and David Ropa, from Ropa Science Research, for providing useful data and correspondence while I was researching anti-oxidant levels in honey. Mike Thurlow, of Orchid Apiaries, has taught me much of what I know about beekeeping. I would like to thank him for sharing his knowledge with me and his bees for making the honey and beeswax I use in my products. I would like to thank Scott Granger, Charted Chemist of Cosmetic Safety Assessment Ltd. for his expertise in product safety and his expertise in chemistry. Finally, I am grateful to Nicole Barnes for reading the manuscript and providing technical help.

Dedication
For George, Nellie,
and Florence

Introduction by Nutritionist Domingo J. Piñero

Honey has been used as food, and for cosmetic and healing purposes for thousands of years, and its different uses have been documented by ancient civilizations and in religious books. The use of honey has always been rooted in tradition and culture, and the effectiveness of its use as folk medicine to treat and prevent different maladies is still based mainly on anecdotal evidence. However, the arrival of commercial beekeeping with the Enlightenment and the posterior improvements to this practice in the mid 1800s led, in the late 19th century, to the beginning of systematic studies of the wound healing and antimicrobial properties of honey. With the recent resurgence of alternative and traditional treatments, the use of remedies containing honey, honey products, and other bee products has also gained new adepts. As a consequence, an increased number of scientific studies seeking to understand the mechanisms of action of these treatments and to identify their active principles have been published in the last several years. Today, it is safe to say that we are starting to have a real understanding of the science of honey remedies and to appreciate the curative value of traditional honey preparations.

Although listing all the uses of honey and honey products preparations would be out of the scope of this introduction, the most common uses are in the treatment of wound healing, ulcers, burns, other skin conditions, infectious diseases, gastrointestinal problems, and allergies. For some of these uses, for example, wound healing and the treatment of burns, the use of honey preparations has been extensively studied in comparison to other standard treatments, with positive results in the majority of the studies. However, the big question for scientists is to understand the possible mechanisms of action for honey in treating (and preventing) disease, and, in recent years, a number of studies have looked into the different properties and curative characteristics of honey and the mechanisms by which those actions are produced.

Honey and honey products have anti-viral properties, anti-inflammatory activity, anti-bacterial capacity, anti-ulcerous properties, and anti-oxidant capacity among other functional characteristics. Some of the beneficial effects of honey are associated with its low pH and high amount of sugar; however, its functional properties are related to the phenolic compounds it contains (not just honey, but also propolis, and royal jelly). Although the concentration of these phenolic compounds in honey varies, depending on a number of factors such as the type of flowers the bees used to create the honey, the season and other environmental conditions, as well as the methods used to process the honey, there is no question that they are the main products associated with the beneficial effects of honey and honey products.

We are just starting to understand some of the specific functions of these phenolic compounds, and in subsequent chapters of this book, Dr. Robb describes in detail how they work, but the simplest way of explaining what they do is that some phenolic compounds in honey are anti-oxidants, and that anti-oxidants contribute to protect us against a number of chronic diseases such as cardiovascular disease, cancer and diabetes. These phenolic compounds also explain the other functions of honey. In vitro, some of these phenolic compounds found in honey are capable of inhibiting viral replication and have anti-viral activity; others help prevent the formation of ulcers, and others inhibit the synthesis of inflammatory compounds, hence giving honey anti-inflammatory activity. The activities of some of these compounds go beyond this limited list, and for some of them we are just beginning to understand their function.

However, we should not think of honey, propolis, and royal jelly just as remedies in the treatment of diseases; they are first and foremost foods. As such, honey was the principal sweetener before the industrialisation of table sugar production. From a nutritional perspective, honey will always come on top when compared to sugar. While sugar is just a source of empty calories, honey has an important role beyond that as a functional food that supplies anti-oxidants in our diet. In the obesogenic environment of our modern day societies, one recommended dietary change that nutritionists agree upon is a reduction in the consumption of added sugar. The recommended daily intake of sugar varies between 24 grams (six teaspoons) for women and 40 grams (ten teaspoons) for men based on standard differences in body size. Although honey is sweeter than sugar, and most people who use honey instead of sugar tend to use smaller amounts, for practical purposes I will assume the use of equal amounts in the following example. If we were to use honey instead of sugar as our sweetener, we would greatly increase our daily intake of anti-oxidants. Just adding honey to the morning cup of coffee will increase the levels of anti-oxidants in 10% of my recommended sugar intake. This is a very modest goal, but one that can significantly increase the yearly intake of anti-oxidants. There are 131 units (FRAP micromoles) of anti-oxidants if you add together the anti-oxidants of 365 cups of coffee with one teaspoon of sugar (a cup of coffee every day for one year). The same 365 cups of coffee with one teaspoon of typical honey give you 4,030 units, and if the type of honey you are using is buckwheat honey, you will get 20,699 units of anti-oxidants. This is an increase of 20,568 units just by using buckwheat honey in one cup of coffee per day instead of sugar. Although no one can say with total certainty the amount of anti-oxidants needed in our diet, we all agree that there is a need to increase the intake of these bioactive compounds. If this increased intake of anti-oxidants attained by using honey instead of table sugar were associated with an increased intake of fruits and vegetables, the other main sources of anti-oxidants in our diets, we would be greatly improving the quality of our diet.

Domingo J Piñero is a professor in the Department of Nutrition, Food Studies, and Public Health of New York University. Research interests in nutritional status of disadvantaged populations, especially obesity and micronutrient deficiencies, and complementary and alternative nutrition.

Chapter 1

Aging, Oxidative Stress and Anti-oxidants

One thing is for certain, we all get older. Our bodies get tired, slow down and do not work as well as they did when we were young. Along with these noticeable changes in ability, there are accompanying changes in appearance. With age comes gray hair and wrinkled skin. Changes in ability and appearance are the result of oxidative stress and damage that accumulates as we age. I suppose you could call oxidative stress the wear and tear associated with living. While there is no way to stop the aging process or bring oxidative stress to a standstill, we can decrease the amount of damage and slow the aging process by increasing the anti-oxidants in our diet and incorporating anti-oxidants into our beauty routines.

Oxidative Stress Causes Aging

Oxidative damage was first recognised, as controlling the aging process, by Denham Harman in 1956. Dr Harman came up with the Free Radical Theory of Aging. In his theory, Harman describes how small, damaging molecules cause cellular damage. Cellular damage accumulates as we live and results in our bodies showing the signs of aging. These small, reactive molecules are known as "free radicals" due to their chemical structure and nature. Free radicals are very reactive molecules which are loose within our cells. Harman described how these loose free radicals could be captured and neutralised by anti-oxidants. Harman's pioneering research showed that anti-oxidant supplements, including vitamin C and vitamin E supplements, decrease oxidative stress and increase life span.

Free radicals are damaging because of their molecular structure. When a free radical is formed, the molecule has an unpaired electron, or a free electron. It is this unpaired electron that makes free radicals so reactive. The free radical bounces around and each time it hits a vulnerable molecule, it damages that molecule. Imagine a pin-ball machine. The pin-ball represents the free radical. The score the number of molecules damaged. Once the free radical is formed, it banks off the sides and hits random spots. Each of the sites the ball hits is a site that is damaged by the free radical. As the free radical bounces around, the score goes up, the damage increases. When the ball goes out of play, that free radical molecule is no longer damaging. One way to remove a free radical is to use an anti-oxidant. An anti-oxidant is a molecule that can react with a free radical and stop it from bouncing around and causing damage.

In our bodies, certain molecules are more vulnerable to damage by free radicals than others. Lipids are very vulnerable, as are proteins and DNA. Cell membranes are made up of lipids, or fats. When cell membranes are attacked by free radicals, the fats are oxidized and the membrane looses fluidity. We are all familiar with the visible effects of aging, our skin gets tired, looses elasticity and develops wrinkles as it is oxidatively damaged. In addition to the physiological changes that accompany aging, there is often decreased physical ability and diminished mental ability. While oxidative damage is the result of biological processes. The reactions which generate free radicals as by-products, are so important, our cells are willing to accept the generation of toxic molecules as a necessary cost.

Production of Free Radicals

Many of the reactions that generate free radicals sustain life, for instance, the reactions that produce energy in cellular respiration. Within our cells, oxygen is converted to energy by the mitochondria. Without cellular respiration, we would die. As oxygen is processed, electrons leak from the mitochondrial respiratory chain and combine with molecular oxygen to form a free radical called superoxide. Reaction I shows the generation of superoxide. The loose electron ($\bullet$) which leaks from the respiratory chain combines with molecular oxygen (O_2) to produce the free radical, superoxide ($O_2^{\bullet-}$).

Reaction I

$$O_2 + \bullet \longrightarrow O_2^{\bullet-}$$

Superoxide bounces around inside the cell until it hits membrane lipids, proteins or DNA, molecules extremely susceptible to oxidative damage. Damage, to lipid membranes, proteins and DNA, results in loss of cellular function, aging and ultimately death of the cell. Our cells have protective mechanisms to eliminate the superoxide radicals produced by oxygen metabolism. These mechanisms limit damage and preserve proper cellular functioning. Superoxide dismutase is an enzyme which removes the superoxide radical by merging two superoxide molecules together. This process is shown in Reaction 2. Two superoxide molecules ($O_2^{\bullet-}$) are merged together by the enzyme, superoxide dismutase (SOD) to form the products, hydrogen peroxide (H_2O_2) and molecular oxygen (O_2).

Reaction 2

$$O_2^{\bullet-} + O_2^{\bullet-} \xrightarrow{SOD} H_2O_2 + O_2$$

The process of eliminating superoxide removes two free radicals but makes hydrogen peroxide as a by-product. Unfortunately, hydrogen peroxide is a very reactive molecule, which can easily be converted to more free radicals. Hydrogen peroxide can react with iron in our cells to produce an even more reactive free radical called hydroxyl radical. Because of the importance of iron in biological functions, iron is available, in our cells, to participate in cellular biochemistry. When cellular iron encounters hydrogen peroxide, the result is production of hydroxyl radical, damage to membranes, proteins and DNA, again resulting in oxidative stress. The reaction that generates free radicals from peroxide is called the Fenton Reaction. Reaction 3 shows the Fenton Reaction. Hydrogen peroxide (H_2O_2) reacts with ferrous iron (Fe^{2+}) to produce the free radical, hydroxyl radical ($^{\bullet}OH^-$) and ferric iron (Fe^{3+}).

Reaction 3

$$H_2O_2 + Fe^{2+} \longrightarrow {}^{\bullet}OH^- + Fe^{3+}$$

The Fenton Reaction describes the reaction of hydrogen peroxide with ferrous iron (iron with a 2^+ charge). The hydrogen peroxide is cleaved by ferrous iron to produce the free radical, hydroxyl radical, and in the process, the ferrous iron is oxidized to ferric iron (iron with a 3^+ charge). Hydroxyl radical is probably the most damaging free radical produced in our bodies. Hydroxyl radical causes the most oxidative damage in our cells and is responsible for aging.

Because hydrogen peroxide is very reactive, our bodies have devised ways to eliminate hydrogen peroxide as it forms. The enzyme, catalase, neutralises hydrogen peroxide by removing it from cells. The neutralising reaction of catalase, is displayed in Reaction 4. Two hydrogen peroxide molecules (H_2O_2) are combined, by the enzyme catalase (CAT), to produce water (H_2O) and molecular oxygen (O_2).

Reaction 4

$$2H_2O_2 \longrightarrow H_2O + O_2$$

Catalase removes the reactive peroxide molecules and produces reaction products, which are less harmful to the cell, water and oxygen. This system works wells, but some hydrogen peroxide molecules escape neutralisation. Additionally, oxygen can begin the oxidative cycle again if an electron escapes the electron transport chain and combines with oxygen to form superoxide as in Reaction 1. Oxidative stress has come full circle, with the product of Reaction 4 able to produce more free radicals, by the process shown in Reaction 1. This is a vicious cycle, perpetuating oxidative damage, but our bodies have other mechanisms to interrupt the cycle and decrease the amount of oxidative damage.

Anti-oxidant Mechanisms in the cell

Generation of free radicals is a normal part of cellular metabolism and the price our cells are willing to pay to generate cellular energy. Throughout our lives, these reactive molecules are formed in small quantities. The generation of free radicals slowly damages our cells and tissues, with the damage manifesting itself as aging. We do not age overnight but see gradual changes. We notice a grey hair or a new wrinkle and feel the wear and tear on our bodies. As previously mentioned, the Fenton Reaction and generation of free radicals through the chemical reaction of hydrogen peroxide and iron is an important mechanism of oxidative stress and aging. While we cannot eliminate the generation of free radicals, because they are by-products of metabolic processes, our bodies have a number of mechanisms in place to minimise the damage. The mechanisms to decrease oxidative stress and aging include; neutralising proteins, iron chelation and anti-oxidant removal of free radicals.

Neutralising Proteins

The first mechanism, our body uses, to reduce oxidative stress is neutralising proteins. The reaction of the first neutralising protein, superoxide dismutase, which neutralises superoxide, is shown in Reaction 2. While the toxic molecule superoxide is eliminated by superoxide dismutase, unfortunately a second toxic molecule is formed, hydrogen peroxide. Our cells have a second neutralising protein, which will eliminate hydroxide radical called catalase. As shown in Reaction 4, catalase can remove hydrogen peroxide from our cells before the peroxide has a chance to react with iron and form hydroxyl radical. Catalase chemically combines two peroxide molecules, to make water and oxygen, as products. Water and oxygen are products the cell can use. Neutralising proteins use chemical reactions to neutralise free radicals and by eliminating the reactive oxygen species, prevent these damaging molecules from attacking our cells. However, as mentioned above, some of the superoxide and peroxide can escape these anti-oxidant proteins. Our cells have other mechanisms to decrease production of free radicals and limit oxidative damage to cells, including iron chelation.

Iron Chelation

The second protective mechanism our bodies have against oxidative stress is to decrease the availability of iron in our cells. Iron is absolutely necessary to life and is involved in many vital processes, so iron cannot be completely eliminated. It is pathological to have too little iron because of the role iron plays in physiology. Indeed, if a person does not have enough iron, they develop a condition called anaemia. Many of us are familiar with this condition and so can easily comprehend that iron is essential for our bodies to function. Iron is necessary for our cells to carry oxygen, used in DNA synthesis and essential to all energy production in our bodies. Iron is essential for life.

Iron is useful to our bodies because it is highly reactive. However, the high reactivity of iron is also responsible for generation of free radicals. In order to limit the reactivity of iron, our bodies have two main proteins which are responsible for controlling the bio-availability of iron. The first, of these proteins, is transferrin. Transferrin is a protein that carries and transports iron to the areas in the body where iron is needed and keeps too much iron from being delivered to locations where iron would be damaging. The second way, our body controls iron's availability, is by storing iron in a protein called ferritin.

Ferritin contains the iron, and keeps iron from reacting in places it should not. Ferritin is like a little vault, which holds the iron molecules and only releases iron when it is needed by the cell. By keeping the iron under lock and key, ferritin can prevent excess iron from reacting with peroxide and generating free radicals. Through this mechanism, ferritin can also help limit oxidative stress.

My thesis research showed that removing iron from cells prevented peroxide-induced oxidative damage, aging and cell death. The iron chelator used in my research was desferrioxamine. Desferrioxamine chelates iron in a very tight manner, making the iron unavailable for biological processes, as well as preventing iron from reacting with hydrogen peroxide. Desferrioxamine prevents the Fenton Reaction and production of hydroxyl radical. While tight chelation of iron sounds like a desirable property, decreasing the availability of intracellular iron leads to anaemia, or iron deficiency. An ideal situation would be chelating iron in a way that iron is available for important biological functions, yet the iron is prevented from reacting with hydrogen peroxide.

Anti-oxidants

While neutralising free radicals and iron chelation help limit free radical formation, there will always be free radicals that escape the first two protective mechanisms and these need to be dealt with by another mechanism. The third way, our cells deal with free radicals, is by using anti-oxidants to eliminate the radicals. This is the mechanism we have the most control over. We cannot control the detoxification enzymes or iron transport and storage, but we can increase our cellular anti-oxidants by taking supplements and increasing dietary anti-oxidant consumption.

Many of us are familiar with vitamin C and vitamin E as cellular anti-oxidants. Both vitamin E and vitamin C are essential nutritional compounds, as they are acquired in our diets and cannot be synthesised in our cells. The amount of vitamin C and vitamin E you have in your cells directly correlates with the amount you consume in your diet. Another very important group of dietary anti-oxidants are the polyphenol anti-oxidants. Polyphenol anti-oxidants are acquired in diet and the physiological concentration depends upon the types of food you eat; whether these foods are high or low in polyphenol anti-oxidants.

Dietary differences explain why people can have different anti-oxidant levels. Anti-oxidant molecules decrease oxidative damage, slow aging and can help prevent disease by eliminated free radicals. Anti-oxidants act like little sponges that soak up free radicals and prevent them from damaging cellular membranes, proteins and DNA. Of the three ways our bodies can decrease oxidative stress and aging, mechanism three is the mechanism we can influence. By increasing our intake of anti-oxidants we can reduce the amount of oxidative damage to our bodies, slow aging and even prevent disease. Dietary anti-oxidants are essential to decrease oxidative damage and aging.

The anti-oxidant polyphenols; flavonoids and phenolic acids, work as anti-oxidants through free radical scavenging and decreasing hydroxyl radical formation. The mechanism, by which polyphenols decrease hydroxyl formation, is likely by binding iron. Polyphenolic molecules have a number of side groups, which can bind iron. The way the polyphenols bind iron is important. The iron bound by polyphenols appears not to react with hydrogen peroxide in the Fenton Reaction. Polyphenols chelate iron in a way which leaves the iron bio-available, for important cellular processes, yet inhibits generation of free radicals. Chelation of iron, by polyphenols, in a way that inhibits the Fenton Reaction, makes polyphenols important biological anti-oxidants, which can significantly reduce oxidative stress and slow aging.

Anti-oxidants Decrease Aging

Today many of us take anti-oxidant supplements to increase our anti-oxidant capacity. While taking supplements can help increase anti-oxidant capacity, we can also use functional foods to increase our anti-oxidant capacity. Eating foods high in anti-oxidants will increase anti-oxidant capacity and protect you from oxidative damage, slow the aging process and will help prevent disease. Polyphenols are important anti-oxidants in foods, including honey. These molecules often add flavour and colour, as well as acting as anti-oxidants. Foods rich in polyphenols are often darker in colour, for instance honey compared to sugar.

One relatively easy way to increase the amount of anti-oxidants in our diets is to substitute honey for sugar. Sugar calories are empty calories, dissociated from vitamins, minerals and anti-oxidants. These calories provide an excellent opportunity, to increase nutritional value, by substituting honey for sugar. Eating honey has been clinically proven to increase anti-oxidants in humans. Honey and other bee products provide us with a plethora of anti-oxidants which can serve as dietary anti-oxidants and anti-oxidants in cosmetic applications. Anti-oxidants can protect skin from oxidative damage and decrease the signs of aging.

Chapter 2

Anti-oxidant Products from the Hive

Some bee products are useful as dietary anti-oxidants, while others are good sources of functional anti-oxidants for beauty products. Products from the hive can be divided into two categories; those physiologically produced by bees and those made by bees with external substances they have collected. Bees synthesise beeswax and royal jelly through biological processes, while bees make bee bread, propolis, and honey, with materials they collect. Bee products contribute to health and beauty through their strong anti-oxidant activity. Particularly, honey, propolis, and royal jelly act as free radical scavengers. The anti-oxidant activity of bee products varies, however all are capable of neutralising hydrogen peroxide, superoxide and hydroxyl radical.

My great-grandfather, George, was a beekeeper and he and his family used bee products in the home, including, beeswax, propolis and honey. People took advantage of the healing properties of bee products. Although at the time my great-grandparents lived, people did not known that bee products contained vitamins, minerals, proteins and anti-oxidants, people did believe, nonetheless, that bee products had healing qualities and were good for you.

Beeswax

Beeswax is physiologically produced by honey bees. Small flakes of wax, are secreted by glands on the abdomen of young bees and collected by worker bees. Worker bees chew the flakes of wax, mixing the wax with saliva, and introduce enzymes. Bees chew the wax until it is malleable. They use the malleable wax to construct the hive's comb. Beeswax is shaped into a strong, hexagonal frame that will hold the bees' honey and young. Beekeepers remove honey from honeycomb, melt the comb and press the beeswax into sheets for the bees to use as a foundation for more honey comb production. In this way, the beeswax is recycled. The harvested comb yields more wax than is needed by the bees. The surplus is used in a number of products for the home.

Beeswax is used to make candles, polishes, in pharmaceuticals, home remedies and to make cosmetics. Beeswax has for a long time been used to make cosmetic preparations such as lotions, creams and moisturising balms because beeswax is an emollient and acts to protect and heal skin. My great-aunt Bea made a number of remedies with bee products. Many of these preparations contained beeswax, for its healing properties as well as used beeswax as a vehicle for other medicinal agents. My relatives did not realise beeswax was also a source of polyphenol anti-oxidants, which facilitate healing and add anti-aging properties, to beeswax cosmetic preparations.

Beeswax is lipophillic, or attracted to oily substances. This quality makes beeswax very easy to mix with other waxes, oils and fats. Beeswax is often mixed with oils and fats to make healing balms. However, due to the hydrophobic nature of beeswax, beeswax cannot be mixed with water without the use of an emulsifying agent. Beeswax blended with water and an emulsifying agent, produces wonderful beauty creams, adding healing properties, nutrients and anti-oxidants.

Royal jelly

Young worker bees secrete royal jelly from their hypo-pharyngeal gland. Royal jelly is an acidic, viscous, aromatic substance with a creamy consistency. It is highly nourishing, fed to young larvae in early development and to the larva intended to be a queen. Because royal jelly is fed to the queen and the queen bee lives significantly longer than the worker bees, royal jelly is associated with increased life span. The queen bee also lays thousands of eggs in her lifetime and is symbol of fertility. Royal jelly supplements are believed to keep you young, help you live longer and increase fertility. Attributing longevity and fertility to royal jelly supplements may be mythical, but there are a number of functional properties that have been identified in royal jelly. Functional properties include; anti-bacterial properties, disinfectant activity, anti-inflammatory properties and anti-oxidant activity. Royal jelly has been shown to scavenge free radicals but royal jelly is not very stable and activity is lost during storage. Even though royal jelly is not very stable, the myth persists that royal jelly keeps you young, increases fertility and promotes long life.

Bee Bread

When honeybees visit flowers to collect nectar, they also collect pollen. Pollen is a proteinaceous, powder produced by flowers. Bees collect pollen as a protein source. Worker bees pack the pollen into balls, sometimes mixed with nectar or honey. This processed pollen is called bee bread. The qualities of bee bread will vary according to botanical source because pollen is unique to the plant which produces it, varying in vitamins, minerals, amino acids and anti-oxidants. Bee bread is sometimes used as a dietary supplement and has been shown to neutralise hydrogen peroxide, superoxide and hydroxyl radical. The high anti-oxidant capacity of bee bread may have cosmetic applications, for instance to decrease oxidative damage to skin and slow the signs of aging.

Propolis

Bees collect plant resins and mix these resins with beeswax to produce the substance called propolis. Of all the bee products, propolis may contain the highest anti-oxidant activity however, the anti-oxidant capacity of propolis can vary greatly. Depending on the plant source of the resin, propolis can vary in colour, aroma, tackiness and in concentration of anti-oxidants. Propolis can be yellow in colour to an almost black colour. Plant resins differ by location and time of year, resulting in regional and seasonal differences in propolis. Propolis contains up to 50% resin and a mix of beeswax, fatty acids and other volatile oils, as well as polyphenol anti-oxidants. Bees use propolis to repair the hive and protect the colony from disease. Honey bees take advantage of the anti-bacterial, anti-microbial and anti-viral activity of propolis and often use the propolis as a hive anti-sceptic. The effectiveness of propolis, to protect the hive, will depend on the source of the plant resin. Propolis samples will have different levels of anti-oxidants, anti-bacterial and anti-microbial activity as the plant source varies, yet all propolis has these qualities.

Propolis samples can vary in colour and fragrance. European propolis, is made with resin primarily from: alder, birch, hazel, oak, poplar, and willow trees, while in the USA, poplar and pine trees are the main source of resin for propolis.

My great grandparents lived on a dairy farm. My great-grandfather, George, would test his propolis by adding it to a fresh cup of milk. The longer the milk remained fresh, the better the propolis crop. Apparently, a good crop of propolis could keep the milk fresh in the summer for a number of days. This is a wonderful demonstration of the anti-bacterial quality of propolis. Scientific research has shown that propolis is active against a number of bacteria including bacillus (the bacteria that causes tuberculosis), staphylococcus, streptococcus, and streptomyces. As well as having anti-bacterial activity, propolis is also anti-fungal, and anti-viral, with activity against herpes and influenza. Finally, propolis has anti-tumour activity and can act as an anti-oxidant. Propolis is an excellent free radical scavenger, able to neutralise hydrogen peroxide, superoxide and hydroxyl radical. The free radical scavenging ability of propolis is a result of a number of anti-oxidants, which originate in the plant resin.

Propolis tinctures are a source of dietary anti-oxidants, used to make home remedies, and used in cosmetics as anti-oxidants. You can also use propolis in a number of preparations, for instance added to soaps, creams or ointments, to control skin conditions such as acne and psoriasis. Propolis cough candies can help sooth sore throats and propolis-containing ointments are useful to treat cold sores, and help prevent infection in cuts and abrasions. Many of the home remedies, which had anecdotal evidence to support their efficacy, now have scientific evidence to support their use. Propolis has great value as an anti-oxidant dietary supplement, in home remedies and in anti-aging beauty products.

Honey

Honey is the product most often associate with bees. Many varieties of honey begin as nectar. Nectar is an aqueous substance produced by flowers, which contains sugars, vitamins, minerals, proteins and anti-oxidants. The chemicals in a flower's nectar will also be present in the honey made with nectar from that flower. Honey bees also make honey with other plant and insect secretions, such as tree sap and honeydew collected from insects. Honey, produced from different sources, also has unique characteristics, including flavour, colour and chemical composition. Honey bees collect the nectar, sap or honeydew, then return to the hive and transfer these fluids to the cells in the honeycomb. The liquid concentrates in the comb into mature, viscous honey. There are countless varieties of honeys created from nectars, honeydews and saps.

Two characteristic that vary in honey from different sources are honey colour and the amount of anti-oxidants. Honey contains polyphenol anti-oxidants, which are pigmented molecules that add colour to honey. The polyphenol anti-oxidants decrease oxidative stress and aging by neutralising free radicals, including, hydrogen peroxide, superoxide and hydroxyl radicals, as well as chelating iron. Eating honey in place of refined sugar can significantly increase dietary vitamins, minerals and most importantly, dietary anti-oxidants. The functional properties of honey can also be useful in home remedies and cosmetic preparations.

Chapter 3

Honey: Nutrition and Anti-oxidants

Carbohydrates, including sugars, are part of a healthy diet and are used by our bodies to produce energy. Sugar is very palatable and as a result, many of us eat more sugar than we should. The United States Department of Agriculture (USDA) recommends that adults limit their daily sugar consumption to 12 teaspoons (48 grams of sugar), for a diet of 2,200 calories, and 18 teaspoons (112 grams of sugar), for a diet of 2,800 calories. Most of us consume much more sugar than the recommended daily amount. The American Heart Association (AHA) reports the average amount of sugar, consumed each day by adults, to be 22 teaspoons (88 grams of sugar), more than double the USDA recommended amount. The AHA also recommends much lower intake levels than the USDA, suggesting women limit daily sugar intake to six teaspoons (24 grams of sugar) and men limit their daily sugar intake to nine teaspoons (36 grams of sugar). For those who consume the average daily amount of sugar, as determined by the AHA, limiting their sugar intake to the guidelines suggested by the USDA may be a more manageable goal than trying to reduce sugar intake to the low levels suggested by the AHA. In addition to limiting sugar consumption, to the recommended daily amounts, replacing sugar with a more nutritious sweeteners will lead to a healthier diet.

Many of us are happy with our diet and not interested in making drastic changes to our eating habits. However, you can continue to eat the same foods, substituting a bit of honey for the sugar you use in your coffee, tea and recipes, to increase nutrition. Sugar contains little in the form of vitamins, minerals and anti-oxidants. Sugar calories are empty calories or calories disassociated from nutritional value. You can increase the vitamins, minerals and anti-oxidants in your diet by choosing a more nutritious sweetener than sugar, such as honey.

Nutritional Comparison of Sweeteners
A major difference, between sugar and honey, is the nutritional composition. The USDA has compiled nutritional data on a variety of food ingredients, including the sweeteners; high fructose corn syrup (HFCS), corn syrup, white sugar (sugar), honey, maple syrup, and molasses. The nutritional composition of sweeteners can be found in the National Nutrient Database for Standard Reference. These USDA standards are useful to make comparisons of the nutritional values.

There are countless varieties of honey; each with a unique set of constituents, such as concentrations of carbohydrates, vitamins, minerals, and anti-oxidants. Table 1 shows the characteristics of the USDA Standard Reference honey sample. The USDA's honey standard is average in composition. In terms of nutritional levels, such as concentrations of minerals and sugar composition, this honey sample is the representative standard. Some honey varieties will have higher values and some honey varieties will have lower values than the USDA Standard Reference, shown in Table 1.

Table 1

Composition of Honey*			
Water	17.10 g	Calcium, Ca	6.00 mg
Protein	0.30 g	Iron, Fe	0.42 mg
Ash	0.20 g	Magnesium, Mg	2.00 mg
Fiber	0.20 g	Phosphorus, P	4.00 mg
Total Fat	0.00 g	Potassium, K	52.00 mg
Fructose	40.94 g	Sodium, Na	4.00 mg
Glucose	35.75 g	Zinc, Zn	0.22 mg
Galactose	3.10 g	Copper, Cu	0.036 mg
Maltose	1.44 g	Fluoride, F	7.00 µg
Sucrose	0.89 g	Selenium, Se	0.80 µg

*Values per 100 grams of Honey

The moisture content of honey varies by plant source, origin and season. The USDA Standard honey sample has water content around 17%. All honey is composed of mostly carbohydrates, or sugars. The types of sugars differ but typically, there is more fructose than glucose and minimal quantities of sucrose. The USDA Standard Reference honey sample contains approximately 40% fructose, 36% glucose and just under 1% sucrose. In addition to these carbohydrates, there are also a number of higher sugars contained in honey and other carbohydrates, such as maltose. Again, the amount of these sugars will depend upon the honey variety. Honey contains a number of minerals, as well as nitrogen sources, such as proteins and amino acids. There are other molecules, which contribute to the flavour and colour of honey, including pigment molecules and aromatic molecules. Most importantly, honey contains anti-oxidants, vitamins and minerals, which are lacking in sugar.

Mineral Content
The mineral content of HFCS, corn syrup, sugar, honey, maple syrup, and molasses, based on the values listed in the USDA Standard Reference data, is presented in Table 2.

Table 2

Mineral Content mg per 100g Sweetener			
	Calcium	Iron	Potassium	Selenium
HFCS	0.00	0.03	0.00	0.0007
Corn Syrup	13.00	0.00	1.00	0.0007
Sugar	1.00	0.05	2.00	0.0006
Honey	6.00	0.42	52.00	0.0008
Maple Syrup	102.00	0.11	212.00	0.0006
Molasses	205.00	4.72	1464.00	0.0178

Upon examining the USDA data in Table 2, comparing the mineral content of each of the sweeteners, it becomes apparent that molasses comes out on top, with the highest levels of calcium, iron, potassium and selenium. Maple syrup contains the next highest levels of calcium and potassium and honey comes in second in the content of iron and selenium. The lowest levels of calcium and potassium were found in HFCS and the lowest level of iron was seen in corn syrup. Compared to sugar; calcium, iron, and potassium are all higher in honey, molasses and maple syrup, with honey and molasses also being higher in selenium.

From this data, it can be recommended that some of your dietary sugar could be substituted with molasses, honey or maple syrup to increase dietary intake of minerals. While molasses contains significantly higher levels of minerals, molasses has a very characteristic, robust flavour, which may not be suited to some dishes. A less strongly flavoured sweetener, like honey, will suit these recipes better. Honey is higher in minerals than sugar and is a more versatile sweetener than molasses, making honey an excellent choice to increase the minerals in a recipe.

Vitamin Content
The vitamin content of different sweeteners used in cooking; HFCS, corn syrup, sugar, honey, maple syrup and molasses, has also been compared. Table 3 shows the values of vitamin C, riboflavin, vitamin B6, niacin and folate, as reported by the USDA Standard Reference data for sweeteners. The highest values, of each vitamin, are highlighted in Table 3.

Table 3

Vitamin Content mg per 100g Sweetener					
	Vitamin C	Riboflavin	Vitamin B6	Niacin	Folate
HFCS	0.000	0.019	0.000	0.000	0.000
Corn Syrup	0.000	0.000	0.000	0.000	0.000
Sugar	0.000	0.019	0.000	0.000	0.000
Honey	0.500	0.038	0.024	0.121	0.002
Maple Syrup	0.000	1.270	0.002	0.081	0.000
Molasses	0.000	0.002	0.670	0.930	0.000

It becomes immediately apparent, from Table 3 that HFCS, corn syrup and sugar are lacking in vitamins. Honey is the only sweetener, which contains vitamin C and folate. Maple syrup contains the highest levels of riboflavin and molasses the highest levels of vitamin B6 and niacin. Honey contains the second highest levels of vitamin B6 and niacin. When comparing the vitamin content of the sweeteners, Table 3 shows honey contains a wider variety of vitamins than HFCS, corn syrup, sugar, maple syrup and molasses. Honey can be recommended as the sweetener of choice, due to its wide range of vitamins.

Protein
Comparing HFCS, corn syrup, sugar, honey, maple syrup and molasses, only honey and maple syrup contain protein. Honey contains 0.3 grams of protein per 100 grams, while maple syrup contains only 0.04 grams of protein per 100 grams. Honey contains 7.5 times more protein than maple sugar. Bees add protein to honey. In order to process nectar, sap and honeydew into "ripe" or concentrated honey, bees add protein enzymes and these remain in the ripened honey. In addition to the protein honey contains, honey is also a source of amino acids. The proteins and amino acids, in honey, also add health benefits to honey not found in the other sweeteners.

Nutritional Summary
To summarise the nutritional value of the sweeteners based upon the USDA's profiles, honey is the only sweetener with vitamin C and folate. Honey also contains proteins and amino acids, a variety of vitamins and minerals. Substituting honey for sugar, you can increase the amount of protein, amino acids, vitamins and minerals in your diet. Molasses and maple syrup also contain a number of vitamins and minerals, however, their characteristic flavours are not always desirable. Honey varieties vary in flavour, making honey a much more versatile sweetener with higher levels of vitamins, minerals and protein than sugar.

Anti-oxidants in Food & Sweeteners

Foods rich in anti-oxidants are part of a healthy diet and can help prevent disease and aging by decreasing the amount of oxidative damage to our bodies. Many of the anti-oxidants found in food are polyphenols. The anti-oxidant polyphenols include flavonoids and phenolic acids. Favonoids and phenolic acids are responsible for anti-oxidant capacity in tea, fruits, vegetables and honey. Polyphenol anti-oxidants act as anti-oxidants by scavenging reactive oxygen species, such as hydroxyl and peroxyl radicals and by chelating iron, and as a result, preventing Fenton production of free radicals. The anti-oxidants in honey add further nutritional value, making honey a functional food, something not applicable to sugar.

Anti-oxidant Capacity

Anti-oxidant capacity can be measured in food samples to compare anti-oxidant levels. Increasing your dietary intake of foods rich in anti-oxidants will increase your physiological levels of anti-oxidants and directly result in decreasing oxidative stress. To take advantage of these anti-oxidants, you needs to know which foods are higher in anti-oxidants. There are two main methods used to determine the anti-oxidant capacity of various food sources, including honey. The first method is the Oxygen Radical Absorbance Capacity (ORAC) method and the second method, a modification of the Ferric Reducing Ability of Plasma (FRAP) method.

Oxygen Radical Absorbance Capacity (ORAC)

The Oxygen Radical Absorbance Capacity of foods is measured using a chemical assay. Much of the oxidative stress, in our bodies is the result of generation of hydroxyl radical from hydrogen peroxide. The ORAC assay measures how well a food inhibits damage, by free radicals generated from hydrogen peroxide. The assay measures the loss of fluorescence of fluoroscein by subjecting it to hydroxyl radical. There is bright yellow fluorescence at the beginning of the assay. The amount of yellow decreases as the yellow probe is damaged by free radicals. Honey is added and the amount of damage to the fluoroscein measured. If honey is high in anti-oxidants, the damage to the probe is prevented and the yellow will not decrease. The brighter the yellow colour, the more anti-oxidants in the honey sample. Less yellow colour indicates lower levels of anti-oxidants in the honey sample.

The ability of the honey sample to decrease the rate of fluorescence quenching of fluoroscein is compared to the baseline and the Trolox® standard. Results of the ORAC assay are reported as micromoles of Trolox® equivalents (TE) per 100 grams of sample (micromoles of TE/100 g). The USDA has a very substantial data set reported in Oxygen Radical Absorbance Capacity (ORAC) of Selected Foods – 2007. This data set was prepared by four agencies; Nutrient Data Laboratory, Beltsville Human Nutrition Research Centre (BHNRC), the Agricultural Research Service (ARS) and the USDA.

The USDA data set is organised alphabetically and included fruits, vegetables, and makes comparisons between raw and cooked samples of the same foods.

Ferric Reducing Ability of Plasma (FRAP)

A second method frequently used to measure the anti-oxidant capacity in biological samples and food samples is the Ferric Reducing Ability of Plasma or FRAP. As indicated by the name of the assay, FRAP was used to measure the levels of anti-oxidants in plasma. Using FRAP, a sample of plasma could be assessed to determine the physiological levels of anti-oxidants present. A modification of this method is useful to measure the anti-oxidant levels in other samples, such as food samples, called Ferric Reducing Anti-oxidant Power. This method determines the levels of anti-oxidant in different food samples, by measuring the increase in blue colour. The amount of colour, is compared to the iron II standards and reported as millimoles Fe^{2+}/100g. A data set which contains a very large sample set of anti-oxidant capacity measured with FRAP was prepared by Carlsen et al (2010) at University of Oslo, Norway. This includes over 3000 food samples measured using FRAP. Samples are divided into 24 categories and include samples of beverages, fruits, vegetables and some sweeteners, including honey.

Comparing ORAC and FRAP Data

While both ORAC and FRAP measure anti-oxidant capacity in food and biological samples, it is difficult to compare measurements made with the two methods. The first obvious difficulty is the standards used to calibrate the two methods differ. Trolox® is used to calibrate ORAC and iron II is used to calibrate FRAP. Additionally, the results of the two assays are reported in different units. ORAC reports anti-oxidant levels in micromoles of TE/100g, while FRAP reports anti-oxidants in millimoles Fe^{2+}/100g. Because the standards and the units are different, you cannot compare samples measured in different data sets.

Although there are differences observed between the data sets, results using the ORAC and FRAP method can be very useful when making dietary choices. Both the USDA and the Carlsen research show similar trends. Both data sets can be used to make dietary choices, which will result in increasing your anti-oxidant capacity, which is a desirable goal.

Anti-oxidant Capacity of Sweeteners

The anti-oxidant capacity of a number of sweeteners are compared using FRAP. The anti-oxidant capacity of corn syrup, sugar, honey, brown sugar, maple syrup and molasses is shown in Figure 1. This data has been converted from millimoles Fe^{2+} / 100g to micromoles Fe^{2+} / 100g to ease comparisons to ORAC data presented later in the text.

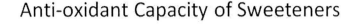

Figure I

Anti-oxidant Capacity of Sweeteners

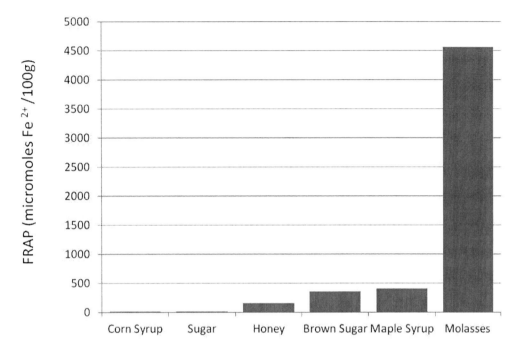

Figure I shows there are large differences between the sweeteners in the Phillips (2009) data set. Corn syrup has the lowest anti-oxidant capacity with only 6 micromoles Fe^{2+}/100g, slightly higher is sugar with 9 micromoles Fe^{2+}/100g. The honey sample used in this study has an anti-oxidant capacity that is 26 times that of sugar (156 micromoles Fe^{2+}/100g). It is worth noting that the honey used in this research was typical, off-the-shelf honey and there are a number of honey varieties with higher anti-oxidant capacities. Brown sugar and maple syrup were much higher than sugar, with anti-oxidant capacities of 361 micromoles Fe^{2+}/100g and 412 micromoles Fe^{2+}/100g, respectively. The highest anti-oxidant capacity was measured in molasses, being over 500 times that of sugar (4562 micromoles Fe^{2+}/100g). Anti-oxidant capacity increases, as the sweetener gets darker in colour. Corn syrup is a clear liquid and has minimal anti-oxidant capacity. Sugar is white and is very low in anti-oxidant capacity. As the colour of the sweetener increases from honey to brown sugar, maple syrup and finally to molasses, the anti-oxidant capacity increases. Molasses is almost black in colour and has a significantly higher anti-oxidant capacity than sugar.

While molasses has a very high anti-oxidant capacity, molasses has a very distinctive taste and is not suitable for some recipes. As mentioned above, the honey sample measured in this study was an average, off the shelf honey. There are a number of other darker honey varieties with higher anti-oxidant capacities. However, even using an average honey in place to sugar will give you 26 times the anti-oxidants capacity of sugar. Many honey varieties have much higher antioxidant capacity, similar to those seen in brown sugar, maple syrup and above.

Anti-oxidant Capacity of Honey

All varieties of honey contain a number of anti-oxidants. The level and profile of the anti-oxidants in honey depend upon the source of the nectar. The anti-oxidant capacity of honey can be taken advantage of simply by increasing your honey intake. Different varieties of honey contain varying amounts of anti-oxidants, so the levels of anti-oxidants are much higher in some varieties of honey, than a typical, off-the-shelf honey. For instance if you substitute ivy, heather or buckwheat honey for some of the sugar in your diet, you will be able to increase your anti-oxidant levels even further. In general, the amount of anti-oxidants in sweeteners increases as the colour increases, this is also true with honey. If you compare honey varieties, very pale honey, such as oilseed rape will have lower levels of anti-oxidants, clover honey, which has a golden colour will have moderate levels of anti-oxidants and very dark honey, such as buckwheat honey will have the highest levels of anti-oxidants. Again, remember that even a very pale honey will have a significantly higher anti-oxidant capacity than sugar.

Figure 2 compares the anti-oxidant capacity of a number of mono-floral honey varieties using FRAP methodology. The samples in Figure 2, were compiled from a number of data sets (Phillips, 2009; Bertoncelj, 2007; Berretta, 2005; and Khalil, 2007). Samples are arranged from low anti-oxidant capacity to high anti-oxidant capacity. For comparison, sugar (sample 1) is included, and an off-the-shelf honey sample (sample 5, shaded black).

Figure 2

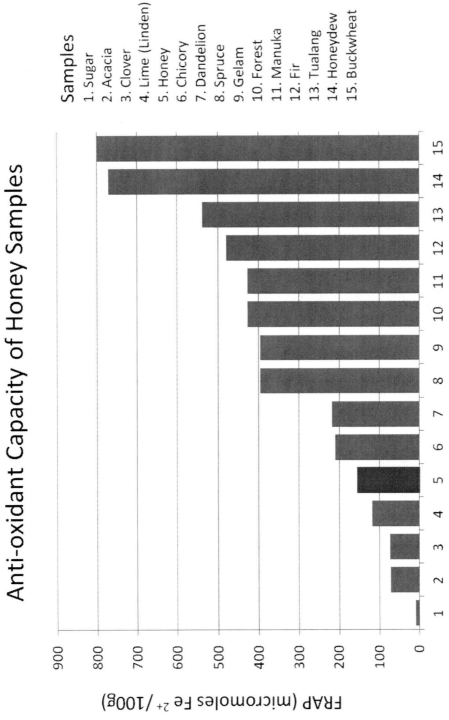

Anti-oxidant Capacity of Honey Samples

Samples
1. Sugar
2. Acacia
3. Clover
4. Lime (Linden)
5. Honey
6. Chicory
7. Dandelion
8. Spruce
9. Gelam
10. Forest
11. Manuka
12. Fir
13. Tualang
14. Honeydew
15. Buckwheat

FRAP (micromoles Fe^{2+}/100g)

The lowest anti-oxidant capacity was measured in the sugar sample. With anti-oxidant capacities nearly eight times the level of sugar, are acacia and clover honeys. Even higher anti-oxidant capacity is seen in lime (linden) honey. Moving up a little are chicory and dandelion honey, with anti-oxidant capacities higher than that seen in the off-the-shelf honey (sample 5). A further increase in anti-oxidant capacity is represented by the manuka and fir honey samples. The highest anti-oxidant capacities were in honeydew honey and buckwheat honey. Interestingly, the anti-oxidant capacity of honey increase as the colour of the honey increases. In Figure 2, the honey with the lightest colour is acacia honey. The colour increases through the honey samples in Figure 2, with buckwheat honey being the darkest in colour. The data presented in Figure 2 illustrates the relationship between anti-oxidant capacity and honey colour. Remember, even the honey samples with the lowest anti-oxidant capacities, have significantly higher anti-oxidant capacities than sugar.

Honey Colour and Anti-oxidant Capacity

There is a relationship between honey colour and anti-oxidant capacity, with lower anti-oxidant capacities found in light-coloured honey and higher anti-oxidant capacities found in dark-coloured honey. This relationship can easily be seen in Figure 2. Just as anti-oxidant capacity increases with sweetener colour (Figure 1), the relationship between colour and anti-oxidant level holds true for honey as well. When choosing a jar of honey, the colour of honey estimates the levels of anti-oxidants. Choosing a darker honey, such as tualang, honeydew or buckwheat honey, can significantly increase your dietary intake of anti-oxidants.

Figure 2 contains only a few of the world's honey varieties. You can use colour to select honey varieties, not included in the chart. If you do not have access the honey varieties in the chart, choose the darkest honey you can find and you will have a honey with high anti-oxidant capacity. In the UK, we have heather and ivy honey, which would work very well as a high anti-oxidant choices.

A large data set which measured a number of characteristics of honey, including the colour and nutritional values of honey, was compiled by Dr Jonathan White (see Appendix 2). The colour of honey historically was determined scientifically by measuring the optical density of honey, using the Pfund scale The Pfund colour grader is an apparatus that uses an amber coloured glass wedge to measure the honey colour. Honey was placed in a wedge shaped cell next to the amber glass wedge and light iused to determine the honey sample's colour measurement. The density measurements were used to classify honey on a colour scale from Water White to Dark Amber. Colour intensity was measured in millimetres, corresponding to the position along the amber wedge of glass and expressed between 1 and 140 mm. The thinnest part of the wedge is lightest in colour, and the thickest part of the amber wedge, the darkest in colour. White divided his honey samples into 13 colour categories (see colour scale on back cover). Today, the USDA classifies honey into seven categories, from Water White to Dark Amber.

Table 4 gives some examples of honey classified using the seven categories. The lightest honey is Water White, followed by Extra White, White, Extra Light Amber, Light Amber, Amber and Dark Amber, with Dark Amber honey being the darkest honey measured on the honey colour scale.

Table 4

Examples of Honey Colours	
Water White	Acacia and Clover
Extra White	Lime (Linden)
White	Honey (average)
Extra Light Amber	Chicory and Dandelion
Light Amber	Spruce, Gelam and Forest
Amber	Manuka, Fir and Tualang
Dark Amber	Honeydew and Buckwheat

Table 4 lists examples of honey colour and their botanical sources. Acacia and clover honeys are very often Water White. Extra White honey is represented by lime (linden) honey. Included for reference is an average honey sample, which is White. A little darker are the Extra Light Amber coloured honeys, chicory and dandelion. Light Amber honeys are spruce, gelam and forest honey. Amber honeys are darker yet and include manuka, chestnut and tualang honey Dark Amber honey includes honeydew and buckwheat honey. While the honey varieties are representative of honey colours, these are by no means, fixed categories. The colour of honey from a botanical source can vary by climate, and region. For instance, buckwheat honey will vary in colour by region, being darker in one area than in another. Additionally, each year the weather conditions may modify the honey crop collected on the same site, resulting in changes in colour, taste, and composition of the honey from the same apiary. Plants produce different quantities and quality of nectar in different climates, the same species of plant in different parts of the world may yield slightly different honey. There will still be trends in honey from the same plant species; for instance, buckwheat honey has a darker colour, stronger flavour and higher levels of anti-oxidants than those in acacia honey, which is lighter in colour.

Polyphenol Anti-oxidants

One type of anti-oxidant associated with honey is a group of molecules called polyphenols, which include phenolic acids and flavonoids. The phenolic content of honey represents the amount of anti-oxidants in honey samples. The total phenolic content differs from the anti-oxidant capacity. The phenolic content measures anti-oxidant amount, while the anti-oxidant capacity measures anti-oxidant activity. The phenolic content measures the levels of phenolic anti-oxidants in honey samples but does not account for the other anti-oxidants present, such as vitamin C and vitamin E. However, the phenolic content can be used to estimate the levels of total anti-oxidants in a sample. The levels of polyphenols in honey vary by honey type and are measured in ranges of 50-800 milligrams per kilogram (mg/kg) honey. Figure 3 shows the phenolic content of a number of honey samples.

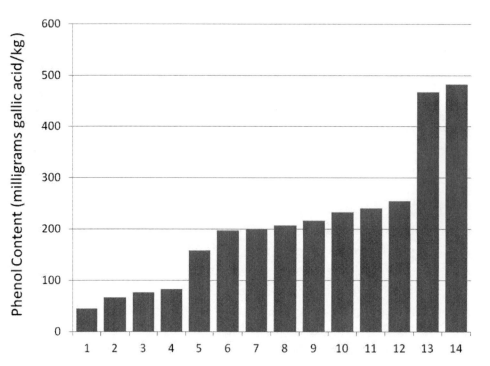

Honey Phenolic Anti-oxidants

Samples
1. Acacia
2. Clover
3. Dandelion
4. Lime
5. Chicory
6. Manuka
7. Heather
8. Jarrah
9. Spruce
10. Forest
11. Fir
12. Honeyde
13. Galem
14. Buckwhe

The honey samples, in Figure 3, are arranged by increasing phenol content. Samples in Figure 3 were compiled from a number of data sets (Gheldof, 2002; Berretta, 2005; Bertoncelj, 2007; Kaškoniene, 2009; Khalil, 2011; Dragar, in preparation). Anti-oxidant levels, measured by phenol content, also increase as honey colour progresses from Water White to Dark Amber. Acacia honey has a phenol content of 46 milligrams gallic acid/kg and is Water White in colour. The phenolic content increases as honey colour darkens to Dark Amber, with buckwheat honey having the highest anti-oxidant level with 796 milligrams gallic acid/kg. If you compare the order of the honey varieties in Figure 2 measured by FRAP, with the phenolic content presented in Figure 3, you will notice there is a slight variation in the order of the honey varieties. In general, both FRAP and phenol content increase with honey colour, but the correlation is not perfect. Nevertheless, you can use honey colour as a reliable predictor of anti-oxidant levels in honey.

Free Radical Scavenging

Another functional way to assess the levels of anti-oxidants in honey is to measure different honey samples ability to scavenge and eliminate free radicals. A method frequently used to assess functional anti-oxidant content is the DPPH radical scavenging method. This method measures a colour change, the greater the accumulation of yellow colour, the better the free radical scavenging activity of the anti-oxidant.

The method for assessing free radical scavenging activity of honey is as follows. A know amount of DPPH radical, which is purple is added. Next a know amount of honey is added and the colour change from purple to yellow is measured as DPPH radical is neutralised (free radical scavenging). The results of the DPPH assay are expressed in Inhibitory Concentration 50% in milligrams per millilitre (IC_{50} mg/ml). The IC_{50} refers to the honey concentration necessary to decrease the DPPH to half of the original concentration, or the amount of honey necessary to scavenge half of the DPPH radical. A higher IC_{50} indicates that more honey is needed to neutralise the radical or DPPH, and the honey sample is low in anti-oxidants. A honey sample with high anti-oxidants will require less honey to neutralise the free radical and as a result have a lower IC_{50}. Honey samples with low IC_{50} values have high levels of anti-oxidants and honey samples with high IC_{50} values have low levels of anti-oxidants. Figure 4 compares a number of honey samples using the DPPH free radical scavenging assay and the results reported in IC_{50} mg/ml.

Figure 4

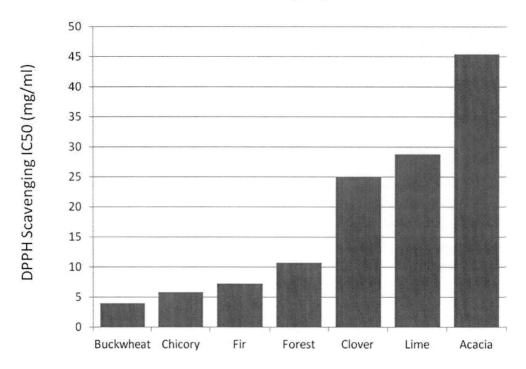

Radical Scavenging of Honey

Figure 4 compares the free radical scavenging ability of a variety of honeys including; buckwheat, chicory, fir, forest, clover, lime, and acacia honey (Beretta: 2005, Bertoncelj: 2007). Using the DPPH method, to assess the free radical scavenging ability, buckwheat honey has the lowest IC_{50}, and acacia honey has the highest IC_{50}. Remembering that lower IC_{50} values indicates higher levels of anti-oxidants. In the honey samples compared, anti-oxidants are lowest in acacia, followed by lime, clover, forest, fir, chicory and the anti-oxidant levels are the highest in buckwheat honey. Again, the anti-oxidant levels, measured by DPPH, increase in honey samples as the colour of the honey sample increases. Lighter coloured honey has less radical scavenging activity and darker honey has higher radical scavenging activity.

Anti-oxidant Levels Consumed per Day
The colour of honey correlates with the antioxidant capacity, amount of anti-oxidants (phenol content) and the anti-oxidant activity (DPPD) in honey varieties. Even if you do not have the scientific data on the levels of anti-oxidants in honey, you can use honey

colour to help you choose a honey that is high in anti-oxidants. An off-the-shelf honey will have higher anti-oxidants than sugar and corn syrup and choosing a dark coloured honey will increase your anti-oxidant levels even further. Table 5 shows the anti-oxidant capacity of; corn syrup, sugar, off-the-shelf honey and buckwheat honey.

Table 5

Sweetener	Anti-oxidant Units 36 g of sweetener	Anti-oxidant Units 72 g of sweetener	Anti-oxidant Units 88 g of sweetener
Corn Syrup	2.16	4.32	5.28
Sugar	3.24	6.48	7.92
Honey	56.16	112.32	137.28
Buckwheat Honey	288.36	576.72	704.88

Units = FRAP micromoles Fe^{2+}/100g

Table 5 compares the anti-oxidant units of sweeteners. The anti-oxidant units found in each sweetener based upon the USDA daily, recommended amount of sugar for a diet of 2,200 calories, is in the first data column. Sugar contains 3.24 units of anti-oxidant, while off-the-shelf honey contains 56.16 units and buckwheat honey contains 288.36 units. Honey contains over 50 antioxidant units more than sugar and buckwheat honey contains more than 285 units higher than the same quantity of sugar. The anti-oxidant units, based upon the USDA daily, recommended amount of sugar for a diet of 2,800 calories, is shown in the second data column. Honey contains over 100 units more than the same quantity of sugar and buckwheat honey contains over 500 units more than the same quantity of sugar. The third data column shows the anti-oxidant units based upon the levels the AHA found adults to consume. In the average amount of sugar consumed per day, there are slightly less than 8 units of antioxidants. Off-the-shelf honey contains 130 more anti-oxidant units. Buckwheat honey contains almost 700 more anti-oxidant units than the same amount of sugar.

As well as containing lower levels of anti-oxidants, sugar also contains little nutritional value, in the form of vitamins and minerals. Sugar calories, are calories disassociated from nutritional value while honey has significantly more vitamins, minerals and anti-oxidants and makes a much better dietary choice. Replacing the sugar in your diet with an off-the -shelf brand of honey, similar to the USDA Standard Reference honey, would increase your anti-oxidant intake by 17 times, compared to sugar. Even more impressive, if you substituted the sugar in your diet with buckwheat honey, the honey with the highest anti-oxidants, your anti-oxidant intake would be 89 times the amount in sugar.

Studies have shown the consumption of honey leads to increased anti-oxidant capacity. Consuming honey also increases blood vitamin E levels and decreases susceptibility to lipid oxidation, a measure of susceptibility to oxidative stress. Beekeepers have a lower biological age than age matched controls with similar exercise patterns. These beekeepers consumed approximately 60 grams of honey per day. The study did not indicate these beekeepers only ate dark honey. It is very likely they consumed their own honey and the consumed honey probably represented a number of honey colours. This study demonstrates that eating a variety of honeys may lead to decreasing the signs of aging. When you compare the anti-oxidant units in each sweetener by daily consumption, you can see how switching from sugar to honey can significantly increase your levels of anti-oxidants. Most of us are not going to replace all of the sugar in your diet with honey, but for each teaspoon of sugar you replace with honey you can increase your physiological anti-oxidants, look and feel younger.

Chapter 4

Cooking with Honey

You can use honey in place of sugar in any drink, food or recipe that requires sweetening. Add honey to a cup of coffee, to savoury recipes and to cookies, desserts and candies. Today, sugar is the most common sweetener used. However, sugar is lacking in nutrients. Food made with honey contains higher anti-oxidants and will increase your anti-oxidant capacity. My great grandparents, Nellie and George, used a combination of honey and sugar in their cooking. While there was a general belief honey was better for you than sugar, people were unaware of the added nutritional and anti-oxidant content in honey.

Years ago, cooks could make substitutions without much thought and frequently did so when ingredients were in short supply. Nellie and George lived on a dairy farm and kept bees. They had a supply of fresh milk, cream, butter and their own honey to use in the food they prepared. One advantage to keeping bees was that George and his family always had honey to use in their cooking. Another advantage was that the food they made with honey, was better for them and very likely helped keep them healthy. Nellie made all her food from scratch and could easily adjust recipes to incorporate the ingredients at hand, including honey. Because so much of what we eat is prepared for us, this is a skill we seem to have lost over the years.

Cooking with honey is not difficult, but there are a few things to keep in mind when using honey instead of sugar. There are five main considerations: the amount of liquid in the recipe, the cooking temperature, flavour, colour, and the amount of acid in the recipe.

Liquid in the Recipe

The most obvious difference between sugar and honey is that sugar is a solid and honey is a liquid. In some recipes, it is very easy to substitute honey for sugar while in others, it is a little more complicated. If you are adding a small volume of sugar, for instance one teaspoon, you can just add honey instead of sugar. If you are sweetening tea, coffee or yoghurt, you can add a spoonful of honey instead of a spoonful of sugar. It is a little more difficult to substitute a larger volume of sugar with honey. This is because the amount of liquid in the recipe changes when you use honey instead of sugar. While honey is a liquid, it is actually composed of water, into which solids are dissolved. In fact, all liquid sweeteners, honey, as well as corn syrup, maple syrup and molasses, consist of solids dissolved in water. You can do an experiment to demonstrate this. Make your own sugar solution by dissolving sugar crystals in water. The solution looks like water but has sugar solids dissolved in the solution. You can see the solids in honey when honey crystallises, or becomes set. When honey sets, the solids come out of solution and the honey becomes more solid looking. Because honey is a solution with solids dissolved in water, you must consider only the amount of water you are adding when you substitute honey for sugar.

When you add ½ cup (120 ml) of honey, you are not adding ½ cup water. Honey is about 17% water, the other 83% is made up of the dissolved solids. The actual amount of water added to a recipe when you use ½ C honey, is approximately four teaspoons (20.4 ml).

The amount of liquid added may be trivial in some recipes, but more significant in others. You need to think about the proportion of sweetener in the recipe. If you are substituting a very small amount of honey for sugar, in a recipe where the other ingredients make up the majority of the volume, the increase in water content will be negligible. However, if the honey makes up a significant volume of the recipe, the change in liquid content may be significant and you may need to adjust to the recipe. You can achieve this in two ways. The first way, is to decrease the amount of the other liquids in the recipe, milk or water for instance. The second way, is to increase the amount of solids in the recipe slightly, for instance add a little more flour.

The percentage of water, in each sweetener, is show in the Figure 5 below and can be used as a guide. Figure 5 compares the amount of water in; sugar, honey, corn syrup, molasses, HFCS and maple syrup.

Figure 5

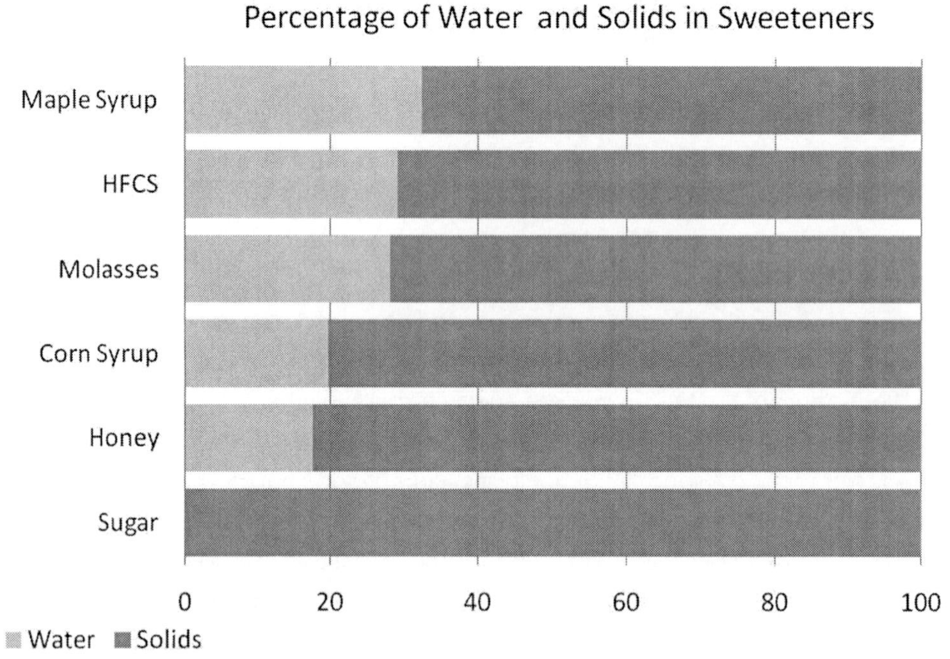

Percentage of Water and Solids in Sweeteners

Water   Solids

Data from *Honey: Its Utilization in Bakery Products.* 1990.
National Honey Board & American Institute of Baking

Sugar is a solid and contains almost no water. Honey contains approximately 17% water. The amount of water increases in corn syrup (19% water), molasses (28% water),

HFCS (29% water) and is the highest in maple syrup (32% water). While all of these sweeteners are liquid, the percentage of water varies. Looking at Figure 5 you can see it is very easy to substitute honey for corn syrup in recipes without adjusting the liquid because both contain approximately 20% water. The amount of liquid in HFCS and molasses are very similar, both with close to 30%. If you would like to use honey instead of molasses or HFCS, there is approximately a 10% difference in liquid in these sweeteners and the recipe may work better if you adjust the recipe to compensate for the 10% less water in honey. Again, you may not need to make an adjustment if the batter feels like it is a proper consistency because the honey is mixed with a number of other ingredients.

If the amount of liquid in a recipe does need to be adjusted, to accommodate the water in honey, in general, the adjustments will not need to be large. The sweetener is only one ingredient and makes up only a proportion of the total volume in a recipe, so the water in honey is into the total volume of the recipe. An extra little bit of flour may be all that is needed in a cookie recipe for instance.

Cooking Temperature

The second consideration, when using honey in recipes, is the cooking temperature. The solid portion in each of the sweeteners is made of different forms of carbohydrates, or sugars. Each of the different carbohydrates has a distinct burning temperature and because of this, recipes using different sweeteners require different cooking temperatures. The sugar we use most often in cooking is sucrose. You can buy sucrose as caster sugar, granulated sugar, light brown sugar and dark brown sugar. Honey has a very different carbohydrate make up containing a combination of fructose and glucose, with a very small percentage of sucrose. The composition of sweeteners is important when considering cooking temperature. Sucrose is a disaccharide, which has a higher burning temperature than the mono-saccharides glucose and fructose. This means a recipe using sucrose can tolerate higher temperatures than a recipe containing a sweetener, which contains either glucose or fructose, such as honey. Because recipes containing honey have lower burning temperatures, than recipes made with sugar, you should use lower cooking temperatures for recipes containing honey to prevent burning.

Maple syrup contains the next highest amount of sucrose with approximately 96% sucrose and only 3% glucose and 1% fructose. Recipes using maple syrup should be cooked at similar temperatures to those containing sugar. If you substitute honey for maple syrup, you should lower the cooking temperature to avoid burning. Molasses is a mixture of approximately 53% sucrose, 21% glucose and 23% fructose. While both molasses and honey contain sucrose, glucose and fructose, the percentages are very different. A typical honey contains only about 1% sucrose, approximately 40% fructose and 36% glucose. The higher percentage of sucrose in molasses means recipes containing molasses can tolerate higher temperatures than recipes that use honey. Again, when honey is used in place of molasses, the temperature will need to be reduced to prevent burning.

The general rule is that recipes containing honey should be cooked at lower temperatures than recipes using any of the other sweeteners. When I am making a recipe with honey, I tend to use a cool oven and reduce temperatures in recipes using other sweeteners between 25° F and 50° F, a reduction in Celsius of as much as 25° C. For cookies, I tend to reduce the temperature much more than if I am making other food with honey. This is in part due to the browning effect associated with using honey instead of sugar. Honey cookies will brown much faster at higher temperatures, so these recipes are cooked at the lowest temperatures.

Colour and Sweeteners

Different sweeteners change the colour of food. Sometimes colour will not be important, while other times changing the colour may make the food less appealing. Some cookies and biscuits we would expect to be pale in colour, such as shortbread or sugar cookies, while other food we expect to be a dark colour such as gingerbread and baked beans. Sweeteners add colour in two ways. The first is simply through the colour of the sweetener itself. Sugar is white and adds little colour, honey can add different amounts of colour depending on the variety, ranging from as white as sugar, to dark amber, close to the colour of molasses. Brown sugar, and maple syrup impart a light brown colour to recipes, while molasses can darken a recipe to an almost black colour.

The second way sweeteners influence the colour of a food is by changing the colour of the recipe as it is heated. We are all familiar with the process. When bread is made, the dough is white. As bread bakes in the oven, the crust becomes a golden brown. The colour change associated with baking is a result of a process called the Maillard reaction, which is the process of non-enzymatic browning as amino acids and reducing sugars react chemically during cooking. The amount of heat can influence the rate of browning, so by lowering the cooking temperature the amount of browning can be somewhat decreased. Reducing sugars involved in the Maillard process include glucose and fructose, while sucrose is a non-reducing sugar and does not contribute to the Maillard reaction and browning by this method. When a recipe uses sugar as the main sweetener, there will be less colour produced by the Maillard reaction than when a sweetener is used which contains glucose and fructose, such as molasses and honey. Amino acids also contribute to the browning associated with the Malliard reaction and honey also contains amino acids, which will further increase browning in recipes using honey. All honey will increase the amount of browning in a recipe when compared to sucrose. If you would like the product to have a light colour, reducing the temperature will help to decrease the amount of browning associated with honey.

Flavour and Sweeteners

In addition to varying the colour of food, adding honey will also change the flavour of a recipe. Consider the desired flavour when substituting honey for other sweeteners in recipes. White sugar adds sweetness without much other flavour. You will get some variation in flavour by using different varieties of sucrose, with flavour increasing as the colour increases from white, to light to dark brown sugar. Maple syrup gives a colour similar to brown sugar and adds a very distinctive flavour from the maple sap. The strongest flavour is associated with using molasses in cooking. The very strong flavour of molasses is very distinctive, with strong hints of liquorish and caramel. The rich flavours of molasses and maple syrup are suited to many recipes but there will be times when the strong flavours associated with molasses or maple syrup will be undesirable.

For me one of the most versatile sweeteners is honey. You can vary the flavour of a recipe by choosing different varieties of honey. Using a honey that is light in colour, such as clover or oilseed rape honey, will add a delicate floral flavour. The same recipe made with a darker honey, such as ivy, heather or buckwheat honey will have a robust honey flavour. Of course there are no rules and part of the fun is to try a recipe with different kinds honey to achieve a variety of flavours. In general, light honey varieties add a delicate floral flavour while darker honey varieties can add stronger more unique flavours to recipes.

Honey and pH

The pH of honey is approximately 3.9 (pH 7.0 is neutral). The lower the pH, the more acidic. This is about the same pH as strawberries, orange juice and yoghurt. Most of us would not consider the taste of honey to be as acidic as orange juice. You cannot always detect the acidity in honey because the sweetness of the honey masks some of the tartness. Using honey in your recipes can change the pH by adding acidity. Many recipes have other ingredients that also increase acidity. For instance, lemon juice contains citric acid and will decrease the pH of a recipe. It is the relationship between ingredients that is important, and how the acidity affects the other ingredients you use. For example, how the acid reacts with leavening agents.

When using honey in baked goods, the choice of leavening agent will depend on the acidity of the recipe. The two main leavening agents used in baking are baking soda and baking powder. Both lighten baked goods with carbon dioxide, by reacting an acid with a base or alkali. The way baking soda and baking powder work is somewhat different. Baking soda is bicarbonate of soda, while baking powder is a blend of sodium bicarbonate and cream of tartar (potassium hydrogen tartrate). Sodium bicarbonate is an alkali. When baking soda is used in a recipe, the acid source comes from the other ingredients in the recipe. Baking soda reacts with an acidic ingredient, such as honey or lemon juice to make carbon dioxide (CO_2). When using baking powder, the baking powder provides both the acid and the alkali. Baking powder contains the alkali sodium bicarbonate and the acid, potassium hydrogen tartrate. The blend in baking powder provides both the acid and the base necessary to product the CO_2.

In general, recipes that contain acids use baking soda as a leavening agent, while those without a substantial acid source use baking powder. Recipes using honey may work best using baking soda if the recipe contains quite a bit of honey and other acidic ingredients. Recipes with small amounts of honey and little acid can use baking powder. When substituting honey for sugar in recipes, the differences in leavening agents should be kept in mind. It may be desirable to use baking soda instead of baking powder.

Recommendations for Cooking with Honey

You can use honey in any recipe. There are a few suggestions when cooking with honey. First, you may need to slightly reduce the amount of liquid in the recipe or slightly increase the dry ingredients to compensate for the water in honey. Second, reduce the temperature by up to 50° F or up to 25° C, when cooking with honey, to decrease the likelihood of burning and to control browning. Thirdly, choose a honey with a colour that works with the recipe, remembering darker honey adds more colour. A fourth recommendation is to select a honey variety with a flavour that suits the recipe. The honey choice could be as simple as what you feel like at the time you make the recipe. Finally, when choosing a leavening agent, use baking soda when the recipe contains large amounts of honey or honey and other acidic ingredients. These five recommendations will allow you to convert any recipe into a healthy honey recipe.

Chapter 5

Starting your Day with an Anti-oxidant Breakfast

Many of us start our day with a cup of tea or coffee, sweetened with a teaspoon or two of sugar. If you add one teaspoon of honey, instead of sugar, you can increase the anti-oxidants. Honey contains anywhere between eight times and 90 times the amount of anti-oxidants in white sugar, weight for weight. Just adding honey to one cup of coffee every day will increase your dietary levels of anti-oxidants. This is a very modest goal and should be easy to achieve. There are 131 units (FRAP micromoles) of anti-oxidants if you add together the anti-oxidants in 365 cups of coffee with one teaspoon of sugar (a cup of coffee every day for one year). The same 365 cups of coffee with one teaspoon of off-the-shelf honey give you 4,030 units and if you add a teaspoon of buckwheat honey each day, you will get 20,699 units of anti-oxidants in 365 cups of coffee. This is an increase of 20,568 units just by using buckwheat honey in one cup of coffee per day instead of sugar.

Breakfast is a time when we have sweet foods. You may have cereal or porridge for breakfast with a spoonful of sugar.   This is another opportunity to substitute honey for sugar. Other breakfast favourites include coffee cakes, muffins, and granola. All of these can be made using honey in the recipe, resulting in higher dietary anti-oxidants. The following honey recipes can help you start your day with anti-oxidants for breakfast. For delicate floral flavours, try light honeys such as acacia, clover or lime honey. For a deeper honey taste, try using forest, honeydew or autumn blend honey.

Chai
Chai is spiced tea prepared in milk, which originates in India. My mother-in-law taught me how to make this recipe, for authentic Indian Chai. Black tea, cloves, ginger and cinnamon, provide a very high blend of anti-oxidants.

Ingredients	U.S.	Imperial	Metric
Milk	2 C	16 fl oz	480 ml
Black Tea		2 Bags	
Honey	1 T	1 tbsp	15 ml
Fresh Ginger (grated)	½ tsp	½ tsp	2.5 ml
Cinnamon (ground)	¼ tsp	¼ tsp	1.25 ml
Cloves		2 Whole Cloves	
Cardamom Pods		2 Cardamom Pods	

Preparation:
1. Place milk, tea bags, honey, and spices in a saucepan.
2. Stirring occasionally, warm chai mixture on medium-low heat until the milk begins to simmer.
3. Remove from heat serve in teacups.

You can increase the amount of spice you add or simmer the chai longer to enhance the spicy flavour. For a richer chai, use whole milk, for a skinny chai, use skimmed milk. Heather honey works very well in chai, with its sweet, yet strong honey flavour. We use whole milk when we want to have a special, spicy drink before bed. Serves two.

Hot Ginger Milk

Ingredients	U.S.	Imperial	Metric
Milk	2 C	16 fl oz	480 ml
Molasses	1 T	1 tbsp	15 ml
Honey	1 T	1 tbsp	15 ml
Ginger (ground)	¼ tsp	¼ tsp	1.25 ml
Cinnamon (ground)	¼ tsp	¼ tsp	1.25 ml
Cloves (ground)		Pinch	
Whipped Cream		Optional	

Preparation:
1. Place milk, molasses, honey, and spices in a saucepan and stir until molasses and honey are dissolved.
2. Warm mixture on low heat until the ginger milk begins to simmer.
3. Remove from heat and cool as necessary.
4. Serve in a pretty tea cup with whipped cream on the top.

Ginger milk tastes just like gingerbread and is wonderful on a cold, winter's morning. Why not serve ginger milk with gingerbread cookies. Honey, molasses, and cloves give ginger milk high levels of valuable anti-oxidants. Dark honeys, such as ivy, honeydew or buckwheat, compliment the molasses and ginger flavour in the ginger milk. For a low fat version, use skimmed milk and omit the whipped cream. Serves two.

Grandma's Rhubarb Sauce

Ingredients	U.S.	Imperial	Metric
Chopped Rhubarb	3 C	4.4 oz	125 g
Water	¼ C	2 fl oz	60 ml
Honey	²/₃ C	5.3 fl oz	160 ml
Cinnamon	½ tsp	½ tsp	2.5 ml
Salt		pinch	

Preparation:
1. Put chopped rhubarb, water, honey, cinnamon and salt in a saucepan.
2. Cook over low heat until the rhubarb has disintegrated.
3. Simmer approximately 10 minutes.
4. Remove from heat.
5. Chill and serve.

My grandmother used to make big pots of rhubarb sauce, which she canned and kept in their cellar. She served a small bowl at breakfast and at lunch. Rhubarb sauce was one of my favourites when I was a child. I would often eat a bowl of rhubarb sauce for breakfast when I visited my grandparents. I liked how the sauce was slightly sweet, yet still tart from the rhubarb. Honey sweetens the rhubarb and thickens the sauce. The cinnamon is optional. I like cinnamon in my rhubarb sauce because that is how my grandmother made it. You could also use nutmeg or a combination of spices. Lime (linden), blueberry or heather honey works well in this recipe. Approximately four servings.

Gogi Berry & Honey Granola

Granola is a mix of oats, dried fruits, nuts, berries and honey. The mixture is toasted to make a healthy and wholesome breakfast cereal. While the dried fruit and nuts vary, granola almost always contains porridge oats and honey.

Ingredients	U.S.	Imperial	Metric
Rolled Porridge Oats	3 C	9.5 oz	270 g
Sunflower Seeds	½ C	2.3 oz	66 g
Sesame Seeds	¼ C	1.1 oz	31 g
Pecans, Chopped	1 C	4.4 oz	125 g
Dried Gogi Berries	1 C	5.6 oz	160 g
Honey	¾ C	6 fl oz	180 ml
Vegetable Oil	$^1/_3$ C	2.7 fl oz	80 ml
Ground Cinnamon	2 tsp	2 tsp	10 ml
Salt	½ tsp	½ tsp	2.5 ml

Preparation:
1. Preheat oven to 325° F or 170° C.
2. Mix together in a large bowl, oats, sunflower seeds, sesame seeds, pecans, gogi berries and cinnamon.
3. Transfer the mixed dry ingredients to a large baking tray or roasting pan.
4. In a smaller bowl, mix together honey, oil, and salt.
5. Pour the liquid onto the dry ingredients.
6. Gently stir the liquid into the dry ingredients to cover the mix.
7. Bake for approximately 30 minutes, stirring often.
8. Cool completely and store in an air-tight container.

Select a dark honey, for a strong honey flavour to contrast the fruit flavour of the gogi berries. The gogi berries are a high anti-oxidant fruit. If you cannot get gogi berries, use another dried fruit. Dried blackberries, cranberries or blueberries would work well. Try varying the dried fruit, seeds and nuts and spices to make different varieties of granola. When we make granola, we use oily honey varieties such as sunflower honey or lychee honey. Approximately, 12 servings.

Popdoodle Coffee Cake
Popoodle coffee cake is a New England recipe from the 1800s. Very easy to prepare with a light honey and cinnamon flavour, popdoodle coffee cake could be made quickly when guests came for coffee.

Ingredients	U.S.	Imperial	Metric
Brown Sugar	¾ C	3.5 oz	100 g
Honey	¼ C	2 fl oz	60 ml
Egg		1	
Vegetable Shortening	½ C	4 fl oz	120 ml
Milk	1 C	8 fl oz	240 ml
Flour	2 ½ C	11 oz	312 g
Baking Soda	2 tsp	2 tsp	10 ml
Ground Cinnamon	1 tsp	1 tsp	5 ml
Salt	¼ tsp	¼ tsp	1.25 ml

Preparation:
1. Preheat oven to 325° F or 170° C.
2. Cream together shortening, honey and brown sugar.
3. Add egg and stir.
4. In a separate bowl, mix flour, baking soda, cinnamon and salt.
5. Fold dry ingredients into the wet ingredients, alternating with milk.
6. Transfer to a greased, 9 x 13 inch pan.
7. Bake for approximately 25 minutes or until the cake springs back when pressed with fingers.

Popdoodle coffee cake is so quick to prepare. Make a plain popdoodle cake, or try adding ½ C chopped pecans or dried fruit to the cake batter. Sometimes we sprinkle the top of the popdoodle coffee cake with brown sugar and cinnamon. This dresses the cake up a bit if you have company visiting or would like to serve this cake for tea. Try using blossom, orange blossom or apple blossom honey. Approximately 12 servings.

Blueberry Honey Muffins
Wild blueberries grew in the wooded, Catskill Mountains, where my grandmother grew up. My grandmother would go with her sister Jessie to find wild "huckleberries". When Florence and Jessie returned with the huckleberries, they would make muffins with their mother, Nellie.

Ingredients	U.S.	Imperial	Metric
Shortening	$1/_3$ C	2.4 oz	68 g
Sugar	½ C	3.5 oz	100 g
Buckwheat Honey	2 T	4 fl oz	120 ml
Medium Egg		1	
Milk	½ C	4 fl oz	120 ml
Flour	2 C	8.8 oz	250 g
Baking Soda	2 tsp	2 tsp	10 ml
Salt	¼ tsp	¼ tsp	1.25 ml
Fresh Blueberries	½ C	2.6 oz	73 g

Preparation:
1. Preheat oven to 325° F or 170° C.
2. Cream together shortening, sugar and honey.
3. Beat egg and add to the shortening mixture.
4. In a separate bowl, mix flour, salt and baking soda.
5. Alternate adding dry ingredients and milk to the batter stirring after each addition.
6. Gently stir blueberries into the muffin batter.
7. Spoon batter into greased muffin tin filling each muffin cup, approximately half full.
8. Bake for approximately 18-20 minutes or until a toothpick comes out clean.

Nellie, Florence and Jessie would have made blueberry muffins with fresh, whole milk from their cows, an egg from their chicken, George's buckwheat honey and the blueberries they collected in the mountain woods. Imagine how delicious their blueberry muffins tasted, made with their own, home-grown ingredients and anti-oxidant rich, wild huckleberries. Approximately six large muffins.

Chapter 6

Savoury Honey Recipes

For many, cookies and candies are the first things that come to mind when they think about cooking with honey. Honey can also be used in savoury dishes, including starters, snacks, main dishes and side dishes. Any recipe that calls for a teaspoon of sugar, will taste even better if you use honey. Sugar adds only trace anti-oxidants and very few minerals to recipes. Honey, on the other hand, adds vitamins, minerals and bio-available anti-oxidants. What is more, honey adds a subtle sweetness and a depth of flavour, to savoury recipes, which may not be expected.

Honey works very well in recipes using chillis, lemons and other spices. Added to ethnic foods, such as Mexican dishes and Indian curries, honey brings out the flavour of the spices. In spicy, full-flavoured recipes, dark, strong-flavoured honeys work very well. Try ivy, heather or buckwheat honey in bakes beans, salsa or curry. In other savoury recipes, try a lighter honey like clover or oilseed rape. Make some of these tasty honey recipes as part of a plan to increase dietary anti-oxidants.

Low Fat Coleslaw

Ingredients	U.S.	Imperial	Metric
Green Cabbage		½ Large, Chopped	
Red Cabbage		¼ Large, Chopped	
Red Onion		½ Small, Diced	
Carrots		2 Finely Grated	
Low fat Mayonnaise	⅓ C	2.8 oz	80 g
Honey	I T	I tbsp	I5 ml
Grainy Mustard	I tsp	I tsp	5 ml
Vinegar	I tsp	I tsp	5 ml
Salt & Pepper		To Taste	

1. Finely chop green cabbage, red cabbage and red onion. Place in a large bowl.
2. Grate carrots and add them to the bowl. Toss with cabbage and onion.
3. In a small bowl, mix mayonnaise, honey, mustard and vinegar and mix well. Add salt and pepper to taste.
4. Add the mayonnaise mixture to the coleslaw and toss gently.

My great-grandmother is the first in our family that made coleslaw with honey. The tradition has continued through the generations. Nellie would have used fresh cream from their dairy cows and George's buckwheat honey in her coleslaw recipe. Today we use low fat mayonnaise in place of cream and use buckwheat, ivy or heather honey. Honey coleslaw is a must for any summer picnic or side dish on a hot summer day. Serves eight.

Crab Louis Pasta Salad

Louis Salad Dressing	U.S.	Imperial	Metric
Mayonnaise*	1 C	8.4 oz	238 g
Chilli Sauce	¼ C	2 fl oz	60 ml
Worcestershire Sauce	1 tsp	1 tsp	5 ml
Honey	1 tsp	1 tsp	5 ml
Lemon Juice	½ tsp	½ tsp	2.5 ml
Salt & Pepper	To Taste		

Salad Ingredients	U.S.	Imperial	Metric
Corkscrew Pasta, Cooked	4 C	19.8 oz	560 g
Green Bell Pepper, Diced	¼ C	2.5 oz	38 g
Red Onion, Diced	¼ C	2.5 oz	38 g
Frozen Peas	¼ C	1 oz	28 g
White Crab Meat	½ C	8 oz	227 g
Salt & Pepper	To Taste		

*To reduce the amount of fat in the recipe, use low fat or fat free mayonnaise.

1. Cook the pasta, drain and let cool.
2. Prepare the Louis dressing by mixing mayonnaise, chilli sauce, Worcestershire sauce, honey and lemon juice in a bowl. Add salt and pepper to taste.
3. In a large bowl, place cooked, cool pasta, diced red onion and green pepper and frozen peas.
4. Add crab meat to the bowl.
5. Pour the Louis dressing onto the pasta salad and gently toss all ingredients.

Crab Louis salad is traditionally made with crab meat and served with a mayonnaise-chilli dressing. This pasta salad, with honey, is a tasty variation of the traditional salad. The honey and crab meat are well suited partners. The sweetness of the honey and crab contrasts nicely with the sharp taste of the dressing. Try a variation by adding boiled egg or capers to the salad. Serves 8.

Honey Baked Beans

Ingredients	U.S.	Imperial	Metric
Dried Beans	½ pound	8 oz	227 g
Baking Soda	1 tsp	1 tsp	5 ml
Red Onion, Chopped	1 C	5.3 oz	150 g
Red Wine	1½ C	12 fl oz	360 ml
Balsamic Vinegar	¼ C	2 fl oz	60 ml
Brown Sugar	½ C	3.5 oz	100 g
Dark Honey	½ C	4 fl oz	120 ml
Tomato Puree	2 T	2 tbsp	30 ml
Wholegrain Mustard	2 T	2 tbsp	30 ml
Salt	3 tsp	3 tsp	15 ml
Pepper	¼ tsp	¼ tsp	1.25 ml

Preparation:
1. In a large bowl or pot, soak dried beans in enough water to cover beans well.
2. The next day, drain the water from the beans and replace with fresh water and 1 tsp baking soda.
3. Boil in pot for 20 minutes.
4. Preheat oven to 150° C or 300° F.
5. Drain the water from the beans.
6. Chop red onion and place in the bean baking dish.
7. In a bowl, mix red wine, vinegar, brown sugar, honey, tomato puree, mustard, salt and pepper.
8. Pour wine mixture over the beans and onions. Mix.
9. The liquid should cover the beans well. If you need to, add enough extra water to just cover the beans and put the lid on the baking pot.
10. Transfer the beans to the oven.
11. Bake approximately 2 hours.

Although making baked beans from scratch takes quite a long time, there is not much effort involved and the taste of the spicy, sweet honey-baked beans is well worth the wait. Honey baked beans are a must for a summer picnic or barbeque. Anti-oxidants come from the red wine, balsamic vinegar, brown sugar and dark honey. Serve hot or cold. Try using heather, buckwheat or any other dark honey in this recipe. Approximately 18 servings.

Honey Guacamole

Ingredients	U.S.	Imperial	Metric
Avocados		4 Medium, Ripe	
Tomato		I Medium Chopped	
Red Onion		½ Small, Diced	
Green Finger Chillis		2, finely Chopped	
Garlic Clove		I, Minced	
Coriander (Cilantro)		Small Bunch Leaves and Stems, Finely Chopped*	
Honey	I tsp	I tsp	5 ml
Lime Juice		Juice of I Small	
Lemon Juice		Juice of ½ Small	
Ground Cumin	I tsp	I tsp	5 ml
Salt & Pepper		To Taste	

*Approximately 1-2 tablespoons chopped

Preparation
1. Halve the avocados, scoop out the flesh and mash in a bowl.
2. Add tomato, red onion, finely chopped chillis, garlic, and fresh coriander (cilantro) and mix thoroughly.
3. Squeeze in lemon and lime juice. Add cumin and honey to the guacamole mix. Stir.
4. Next, add salt and pepper to taste.
5. Mix the guacamole.

Serve honey guacamole with corn chips or tortillas. Honey, chillis and citrus juice work wonderfully together. The honey cuts the heat of the chillis and the lemon is a traditional compliment to honey. Lime juice adds just the right amount of tartness to the guacamole. Serve honey guacamole at a picnic, party or as a tasty snack. For subtle honey flavour, use a lighter honey. For a more robust honey flavour, use a darker honey. Approximately 12 Servings.

Waggle Salsa

Ingredients	U.S.	Imperial	Metric
Fresh Tomato (chopped)	3 Large, Chopped		
Red Onion (chopped)	1 Medium, Diced		
Green Finger Chillis (minced)	Minced (1 for mild, 3 for hot)		
Garlic Clove	1 Minced		
Coriander (Cilantro)	Small Bunch Chopped*		
Honey	2 T	2 tbsp	30 mL
Lime Juice	Juice of ½ Lime		
Lemon Juice	Juice of ½ Lemon		
Salt & Pepper	To taste		

*Approximately 1-2 tablespoons chopped

Preparation
1. Prepare the tomato, red onion, chillis, garlic, and fresh cilantro and place in a bowl.
2. Squeeze in lemon and lime juice. Add honey.
3. Mix the salsa. Add salt and pepper to taste.

Waggle salsa makes a great party snack. You can add fewer chillis if you would like a mild salsa, or more chillis to add a bit of fire to the salsa. Serve waggle salsa with corn chips or a Mexican meal. Try making some easy cheese cassadias. Spread a soft tortilla with waggle salsa and honey guacamole. Add a layer of cheese and sandwich with a second soft tortilla. Toast in the oven until the cheese is melted and you have an easy dinner or a tasty appetiser. Honeydew or forest honey is very tasty in this recipe. Waggle salsa serves 12.

Chilli Sauce

Ingredients	U.S.	Imperial	Metric
Tomatoes		3 large, Diced	
Red Onion		I small, Diced	
Finger Chillis		10-12 Finely Chopped	
Coriander		Stems and leaves Small Bunch Chopped*	
Garlic		7 Cloves Minced	
Lemon Juice		Juice of I Medium	
Lime Juice		Juice of I Medium	
Water	I C	8 fl oz	240 ml
Olive Oil	2 T	2 tbsp	30 ml
Vinegar	2 T	2 tbsp	30 ml
Honey	I T	I tbsp	15 ml
Cumin	I tsp	I tsp	5 ml
Salt and Pepper		To Taste	

*Approximately 2 tablespoons chopped

Preparation
1. Place olive oil and red onion in a saucepan. Cook over low heat.
2. Stir in garlic, chillis, tomatoes and coriander.
3. Add the remaining ingredients to the saucepan and continue to cook over low heat.
4. Reduce the chilli sauce until thick.
5. Remove from heat and cool.

Use chilli sauce as a condiment or as an ingredient in recipes calling for chilli sauce. Adjust the amount of heat to your taste. Simply decrease or increase the number of chillis in the recipe. Try varying the types of chillis you use as well as the type of honey you use for your own special blend of chilli sauce.

Spicy Lamb Balls

Ingredients	U.S.	Imperial	Metric
Minced (ground) Lamb	1 ½ lb	24 oz	680 g
Yellow Onion, Chopped	1 C	5 oz	150 g
Green Chillis	2-3 Finely Chopped		
Chilli Powder	1 tsp	1 tsp	5 ml
Ground Cumin	3 tsp	3 tsp	15 ml
Ground Turmeric	1 tsp	1 tsp	5 ml
Honey	1 T	1 tbsp	15 ml
Lemon Juice	Juice of 1 Lemon		
Coriander, Stems and Leaves*	Small Bunch, Very Finely Chopped		
Salt and Pepper	To taste		

*Approximately one tablespoons chopped

Preparation
1. Place minced lamb in a bowl. Add chopped onions, chillis, and coriander. Mix together.
2. To the lamb, add honey, lemon juice and spices. Mix well, until spices are blended thoroughly, through the lamb.
3. Cover bowl with film and place in refrigerator for one to two hours.
4. Shape lamb mince into 2" balls and place on a baking tray.
5. Cook under the grill/broiler until brown. Turn the lamb balls as necessary.

Serve lamb balls with rice or chapattis or serve them on sticks as tasty little nibbles. The flavour of lamb blends very well the honey, lemon and chilli. Form the lamb mixture into patties, grill and serve in buns for an appetizing lunch. Try oilseed rape or clover honey in this recipe. Lamb balls are so delicious; you may want to double the recipe. Serves six.

Chicken Skewers

Ingredients	U.S.	Imperial	Metric
Chicken Breast	2 Medium Breast Cut into 2" Cubes		
Green Pepper	Cut into 2" Cubes		
Onion	Cut into 2" Cubes		
Tomato	1 Medium, Chopped		
Red Onion	½ Small, Finely Diced		
Green Chillis	3, Finely Chopped		
Garlic Clove	4, Minced		
Coriander (Cilantro)*	Small Bunch Leaves and Stems, Finely Chopped		
Olive Oil	2 T	2 tbsp	30ml
Honey	1 T	1 tbsp	15ml
Lemon Juice	Juice of 1 Small		
Salt	1tsp	1tsp	5ml

*Approximately one tablespoons chopped

Preparation
1. Begin by making the marinade for the chicken. In a large bowl, place tomato, red onion, chillis, garlic, coriander, olive oil, honey, lemon juice and salt. Mix well.
2. Cut chicken breast into cubes and add to the bowl containing the marinade.
3. Toss the chicken cubes in the marinade. Cover the bowl with cling film and place in refrigerator. Leave the chicken in the marinade a few hours to overnight.
4. Prepare the skewers by alternating pieces of chicken, onion and green pepper.
5. Grill the chicken skewers on the barbeque.

Serve honey chicken skewers with rice or pita bread. The honey, lemon and olive oil marinade gives the chicken a tantalising sweet flavour in contrast to the smoky grilled vegetables. Try other vegetables on the chicken skewers, such as cherry tomatoes or mushrooms. Floral honey, such as orange blossom or apple blossom honey work very well with chicken. Serves four.

Undhiyu Indian Vegetable Casserole

Undhiya is a delicious mixed vegetable casserole from Gujarat, India. Traditionally the vegetables are cooked in an earthen pot above a fire. You can make your own undhiya without an earthen pot, in your oven.

Ingredients	U.S.	Imperial	Metric
Yams	½ lb	8 oz	227 g
Green Banana with Skin	2 Medium Bananas Cut into 1" Sections		
Potato	½ lb	8 oz	227 g
Aubergine (Eggplant)	½ lb	8 oz	227 g
Green Beans	¼ lb	4 oz	114 g
Vegetable Oil	¼ C	4 tbsp	60 ml
Honey	2 T	2 tbsp	30 ml
Lemon Juice	Juice of 1 Large Lemon		
Coconut, Desiccated	¼ C	0.8 oz	19 g
Ginger, Grated	2 tsp	2 tsp	10 ml
Garlic, Crushed	1 tsp	1 tsp	5 ml
Coriander (Cilantro)*	Bunch Leaves and Stems, Finely Chopped		
Turmeric	1 tsp	1 tsp	5 ml
Cumin	2 tsp	2 tsp	10 ml
Red Chilli Powder	1 tsp	1 tsp	5 ml
Salt and Pepper	To Taste		

*Approximately three tablespoons chopped

Preparation:
1. Prepare the green bananas by washing the skin and removing the end and the top stem. Discard any stickers on the bananas.
2. Preheat oven to 170° C or 325° F.
3. Chop yams, bananas, potatoes, and aubergines and place in large casserole.
4. Add green beans to the dish with the vegetables.
5. In a bowl, place vegetable oil, lemon juice, honey, coconut, ginger, garlic, coriander, turmeric, chilli powder and salt and pepper and mix well.
6. Pour oil and spices over vegetables and stir.
7. Cover the casserole, and transfer to the oven to bake.

8. Bake approximately 1 1/2 hours, or until the vegetables are tender.

Undhiyu is one of my favourite Indian dishes from the Gujarat region and is the first Indian dish I learned to cook. The blend of winter vegetables is very comforting on a cold day. The banana, eggplant and yam go so well with the honey and spices. This may be the first recipe you make which includes a green banana with the skin on. The banana is transformed in the baking, coming out with black skin and the flesh becoming very starchy. There is still a slight banana flavour but you might not recognise it if you did not know it was there. You can vary the vegetables and honey for different flavour blends. Serve undhiyu with rice or chapattis. Makes 12 servings.

Chapter 7

Honey Baked Goods & Desserts

Most of us think of sweet food when we think of honey. Baked goods present a wonderful opportunity to increase your anti-oxidant levels by substituting some of low-anti-oxidant containing sugar for the high-anti-oxidant sweetener honey. I am not suggesting you eat more bake goods, however we all enjoy a cookie or dessert now an then. When you do fancy a dessert, choosing one made with honey will provide you with significantly higher dietary anti-oxidants. A recipe containing 1 ½ cups of sugar (300 grams) contains 27 units of anti-oxidants from sugar (micromoles by FRAP). Making the same recipe with one cup sugar (200 grams) and ½ cup off-the-shelf honey (170 grams) gives you 283 units of anti-oxidants (micromoles by FRAP), and 1379 units (micromoles by FRAP) if you use buckwheat honey. Substituting honey for sugar will increase the carbohydrate anti-oxidants up to 50 times that of the same recipe using sugar alone.

Many recipes, from years gone by, included honey as well as sugar. My great-grandmother, Nellie made her baked goods with sugar and the family honey. Honey sugar cookies, were made with wild flower honey, as was shortcake. Nellie used buckwheat honey to make gingerbread cookies and honey brownies. All of the recipes contain higher anti-oxidants than those with sugar alone. My great-grandmother and her family just knew they were delicious.

Honey Sugar Cookies

Ingredients	U.S.	Imperial	Metric
Shortening	⅔ C	4.8 oz	137 g
Sugar	1 C	7 oz	200 g
Honey	½ C	4 fl oz	120 ml
Eggs		2	
Vanilla	2 tsp	2 tsp	10 ml
Flour	3 ¼ C	14.3 oz	406 g
Baking Soda	2 tsp	2 tsp	10 ml
Salt	¼ tsp	¼ tsp	1.25 ml

Preparation:
1. Preheat oven to 325° F or 170° C.
2. Cream together shortening, sugar and honey.
3. Mix in eggs, one at a time.

4. Add vanilla to sugar mixture and blend.
5. In a separate bowl, mix flour, baking powder and salt.
6. Slowly add the flour to the wet ingredients, stirring after each addition.
7. Chill cookie dough for approximately one hour.
8. Roll dough and cut out cookie shapes.
9. Bake cookies until the edges begin to brown*

*Baking time will depend on the size of the cookie. For best results, do not mix different sized cookies on the baking tray

This recipe is incredibly versatile. You can make a variety of shapes for any occasion. Sprinkle the tops with coloured sugar before you bake the cookies to add a little magic, or decorate the cookies after baking with honey butter cream tinted with food colouring. Meggy Jayne and Jasmine enjoy getting out all the coloured sugars, sprinkles and shapes to decorate sugar cookies on rainy afternoons. Sometimes we make little cookie sandwiches using small circular cookies filled with the following honey butter cream, perfect for a tea party. Light honey varieties work best for this recipe, like clover, blossom or lime honey. Makes approximately 40 cookies.

Honey Butter Cream

Ingredients	U.S.	Imperial	Metric
Confectioner's Sugar	1 C	4.2 oz	120 g
Butter	$^2/_3$ C	5.3 oz	151 g
Honey	1 T	1 tbsp	15 ml
Vanilla	2 tsp	2 tsp	10 ml

Preparation:
1. Cream confectioner's sugar (icing sugar) and butter together.
2. Blend in vanilla and honey.
3. Mix butter cream until light and fluffy.
4. Adjust consistency if necessary.

Tint the honey butter cream with food colouring if you wish to make different colours. To vary the flavour of our butter cream, we enjoy trying different honey varieties. Use lighter, mild flavoured honey to compliment honey sugar cookies. For a contrast to the flavour of the cookies, try a darker honey such as honeydew or forest honey. Remember the darker the honey, the higher in anti-oxidants.

Ranger cookies

Ingredients	U.S.	Imperial	Metric
Flour	2 C	8.8 oz	250 g
Baking Soda	1 ½ tsp	1 ½ tsp	7.5 ml
Salt	½ tsp	½ tsp	2.5 ml
Butter	1 C	8 oz	227 g
Sugar	1 C	7 oz	200 g
Brown Sugar, packed	½ C	3.9 oz	109 g
Honey	½ C	4 fl oz	120 ml
Eggs	2		
Vanilla	1 tsp	1 tsp	5 ml
Rolled Oats	1 C	5.5 oz	156 g
Desiccated Coconut	1 C	3 oz	75 g
Rice Cereal	1 C	1.1 oz	32 g

Preparation:
1. Preheat oven to 350° F or 180° C.
2. Mix together in a bowl, flour, baking soda and salt.
3. In a separate bowl, cream butter, brown sugar, sugar and honey.
4. Blend eggs and vanilla into the sugar-butter mixture.
5. Slowly add the flour mixture to the wet ingredients and mix.
6. Stir in oats, coconut and rice cereal.
7. Drop cookie dough in spoonfuls, onto cookie sheet with 1" between cookies.
8. Bake approximately six to eight minutes.

My mother made ranger cookies for us when we were children with clover honey. We would come home from school to the delicious smell of warm honey, coconut and oatmeal. My brother Erich liked ranger cookies so much he would eat them right off the cookie tray. Appropriately, Erich grew up to be a park ranger! Ranger cookies are always a family favourite. Recipe makes approximately 60 cookies.

Honey Gingerbread Cookies

Ingredients	U.S.	Imperial	Metric
Shortening	½ C	3.6 oz	103 g
Brown Sugar, packed	1 C	7.7 oz	217 g
Bicarbonate of Soda	2 tsp	2 tsp	10 ml
Salt	2 tsp	2 tsp	10 ml
Ground Cinnamon	3 tsp	3 tsp	15 ml
Ground Ginger	3 tsp	3 tsp	15 ml
Ground Allspice	2 tsp	2 tsp	10 ml
Ground Cloves	1 tsp	1 tsp	5 ml
Dark Molasses	1 C	8 fl oz	240 ml
Honey	½ C	4 fl oz	120 ml
Water	¼ C	2 fl oz	60 ml
Flour	6 ½ C	28.6 oz	812 g

Preparation:
1. Cream together shortening, brown sugar, baking soda, salt and spices, until light and fluffy.
2. Beat in molasses and honey.
3. Stir in water.
4. Gradually stir in enough flour to make a stiff dough.
5. Shape gingerbread dough into a ball, cover with cling film and refrigerate several hours to overnight.
6. Preheat oven to 325° F or 170° C.
7. Roll dough out in small amounts to a thickness of approximately ¼" and cut into desired shapes.
8. Bake for 8-10 minutes or the edges begin to brown*.

*Cooking time will vary by size and shape of the cookies. Do not mix shapes on the baking tray as different sized cookies will take different lengths of time to bake

We enjoy honey gingerbread throughout the year but particularly enjoy making gingerbread Christmas cookies. The aromatic spices in the gingerbread are sure to put you in a festive mood at Christmas time. My Grandmother made gingerbread cookies, with buckwheat honey, each year to hang on the Christmas tree. Gingerbread boys, stars and tree-shaped cookies were strung on red ribbon and hung on the tree. When visitors came, they were offered a festive cookie. Grandma had a small box with more gingerbread cookies to restock the tree. When we went to my grandparents' at Christmas time, the first thing grandma did was take us to the tree to choose a gingerbread cookie. Makes approximately 60.

Strawberry Shortbread

Shortbread	U.S.	Imperial	Metric
Self rising Flour	2 ½ C	11 oz	313 g
Salt	½ tsp	½ tsp	2.5 ml
Shortening	½ C	3.6 oz	103 g
Milk	¾ C	6 fl oz	180 ml
Honey	2 tsp	2 tsp	10 ml
Vanilla	1 tsp	1 tsp	5 ml

Preparation:
1. Preheat oven to 325° F or 170° C.
2. Mix flour and salt in a bowl.
3. Cut in shortening until the mixture appears mealy.
4. In a separate bowl, mix milk, honey and vanilla.
5. Slowly add the milk mixture to the dry ingredients.
6. Turn out into a 9" round pan and gently pat into the pan.
7. Bake in preheated oven, approximately 20 minutes.
8. Cool shortbread.

Strawberry Sauce	U.S.	Imperial	Metric
Sliced Strawberries	4 C	23.4 oz	664 g
Honey	½ C	4 fl oz	120 ml

Preparation:
1. Wash fresh strawberries and remove stems. Slice and transfer to large bowl.
2. Stir in honey and set aside. The honey will draw liquid from the strawberries and make a nice "sauce".

Honey Whipped Cream

Ingredients	U.S.	Imperial	Metric
Double Cream, Chilled	1 ½ C	12 fl oz	360 ml
Honey	3 T	3 tbsp	45 ml
Vanilla Extract*	1 tsp	1 tsp	5 ml

*Vanilla extract with vanilla seeds is delicious in honey whipped cream

Preparation:
1. Place chilled double cream in a large bowl.
2. Add honey and vanilla.
3. Begin to whip with a mixer.
4. Whip until stiff.

Assemble the Shortbread
1. Cut a piece shortbread and slice in half.
2. Place one slice of shortbread into a bowl.
3. Top shortbread slice with strawberries.
4. Place the other half of the shortbread on top of strawberries.
5. Top the shortbread sandwich with a dollop of honey whipped cream.

One day Meggy Jayne and I were making strawberry shortcake when we had a special visitor. We had made the shortbread, sliced the strawberries and put them into the honey. We were just starting to make the honey whipped cream, when we heard a loud buzzing. It is not unusual for bees to come into the kitchen when we are cooking with honey, so neither Meggy Jayne or I were surprised. I usually gently scoot our little bee friends out of the kitchen and go back to what I am doing. I turned to find the buzzing bee. I was definitely surprised! In the kitchen was the largest, most beautiful bumble bee I had ever seen. She was over 2 inches long, plump and covered in fuzz. The bumble had obviously smelled the honey. I turned in surprise and told Meggy Jayne to look at the plump bubble bee. Meggy's eyes grew large in amazement. She said "oh, she is so beautiful, mama". We both stopped cooking for a few minutes to enjoy our time with our new friend. She looked so soft it was all I could do not to touch her. We needed to get back to the work at hand, so we helped her on her way. I am pleased to know we have such a beautiful and friendly neighbour. Serves 6.

Honey Carrot bars

Ingredients	U.S.	Imperial	Metric
Eggs		2	
Brown Sugar	¾ C	5.8 oz	163 g
Honey	¼ C	2 fl oz	60 ml
Vegetable Oil	½ C	4 fl oz	120ml
Flour	I C	4.4 oz	125 g
Baking Soda	I tsp	I tsp	5 ml
Salt	¼ tsp	¼ tsp	1.25 ml
Cinnamon	3 tsp	3 tsp	15 ml
Nutmeg	I tsp	I tsp	5 ml
Grate Carrots	I ½ C	6 oz	173 g
Chopped Walnuts	¾ C	2.3 oz	65 g
Desicated Coconut	½ C	1.5 oz	43 g

Preparation:
1. Preheat oven to 350° or 180° C.
2. Beat eggs until light and fluffy in a large bowl.
3. Slowly add brown sugar and honey to eggs and continue to beat the mixture.
4. In a separate bowl, mix flour, baking soda, salt, cinnamon and nutmeg thoroughly.
5. Alternate the addition of flour mixture and the vegetable oil to the wet ingredients mixing after each addition.
6. Fold the carrots, walnuts and coconut into the batter.
7. Grease a 9"x 13" pan.
8. Pour batter into the pan. Spread the mixture evenly.
9. Bake approximately 20 minutes.
10. When cool, spread with Cream Cheese Frosting (recipe below).

Cream Cheese Frosting

Ingredients	U.S.	Imperial	Metric
Cream Cheese	1 C	7 oz	198 g
Butter	½ C	4 oz	114 g
Honey	2 T	2 tbsp	30 ml
Confectioner's Sugar	1 ¼ C	5.3 oz	150 g
Vanilla	2 tsp	2 tsp	10 ml
Salt	¼ tsp	¼ tsp	1.25 ml

Preparation:
1. Soften cream cheese and butter by bringing to room temperature.
2. Beat cream cheese and butter together in a large bowl.
3. Add honey to the cream cheese butter mixture.
4. Blend in vanilla and salt.
5. Slowly mix in the powdered sugar.
6. To adjust the consistency, you may add more sugar if too thin or a little milk if too thick.

Wonderfully moist, these are a cross between a bar cookie and a cake. The grated carrots, walnuts and coconut add a nice texture. The flavour is very similar to carrot cake with cream cheese frosting. Honey makes these bars very moist. Try heather, lime or blossom honey in this recipe. Store cream cheese carrot bars in the refrigerator. Makes approximately 18 bars.

Honey Brownies

Ingredients	U.S.	Imperial	Metric
Sugar	¾ C	5.3 oz	150 g
Butter	½ C	4 oz	114 g
Honey	1/3 C	2.7 fl oz	80 ml
Eggs		2	
Vanilla Extract	1 tsp	1 tsp	5 ml
Unsweetened Chocolate*	2 squares	2 oz	57 g
Flour	1 C	4.4 oz	125 g

Baking Soda	I tsp	I tsp	5 ml
Salt	I tsp	I tsp	5 ml

*Use baker's, unsweetened dark chocolate in this recipe. If you cannot get unsweetened dark chocolate, use a dark chocolate with high cocoa solids and no milk if possible.

Preparation:
1. Preheat oven to 325°F or 170°C.
2. Cream together butter and sugar until fluffy.
3. Mix honey and vanilla into creamed mixture.
4. Add eggs one at a time mixing thoroughly.
5. Melt chocolate on bain-marie or in the microwave.
6. Add chocolate to wet ingredients and stir.
7. In a separate bowl, mix flour, baking soda and salt.
8. Gradually fold flour and salt into the creamed mixture.
9. Pour batter into a 8x8 inch greased pan.
10. Bake for approximately 40 minutes.
11. Cool and cut into 2" by 2" squares.

Chewy, warm honey brownies are so delicious. You cannot help feeling a bit guilty when you eat them. With chocolate being high in anti-oxidants and the recipe containing honey, you should not feel too guilty. Try using a dark honey with a strong flavour to compliment the strong flavour of the baker's chocolate. Makes 16 brownies.

If you are going to have a sweet treat or dessert, have a dessert made with honey. A cookie or brownie and a cup of coffee with a teaspoon of honey is a great way to end a meal. Honey brownies and coffee with honey has more anti-oxidants than the same using just sugar. Remember the anti-oxidants in honey will keep you young and slow aging. I am sure you will be tempted into having dessert!

Chapter 8

Confections

When my grandmother, Florence, was a child in the early 1900s, beautiful confections were regularly made at home. All that you need to make candies is sugar, water, a saucepan, stove and a little knowledge of candy making. You can use the ingredients in the larder to make caramel sauce, fudge, toffee, brittle and butterscotch, simply by changing the temperature of the sugar syrup and adding a few other ingredients. Often confections are made with corn syrup, while sweet, these candies are nutritionally lacking. Adding honey to the sugar pot adds minerals, vitamins and most importantly anti-oxidants. I am not recommending you increase your sugar intake, only that when you have a candy, you choose a honey confection. You will benefit from higher levels of anti-oxidants than if you have candies made with sugar and corn syrup. When comparing corn syrup and honey using FRAP, it becomes apparent honey has significantly higher anti-oxidants capacity. In ¼ cup or 60ml of corn syrup, there are 5 anti-oxidant units, in a typical honey variety, there are 133 anti-oxidant units and in buckwheat honey there are 681 anti-oxidant units (units = micromoles), 136 times the anti-oxidant level in corn syrup.

My family has made honey candies for four generations, beginning with my great-grandmother, Nellie. When Nellie made her buckwheat honey fudge, she did not know honey was higher in anti-oxidants than corn syrup, only that it was delicious. When my grandmother was a child, she, and her mother Nellie would prepare a fantastic selection of confections each year at Christmas time. Their candies were particularly special because they contained my great-grandfather, George's honey. Nellie and Florence made caramel sauce and blackberry syrup, with George's wild flower honey and made fudge and peanut brittle with George's buckwheat honey. Often the flavouring was very simple, but the honey candies were a special treat, nevertheless. Nellie's buckwheat honey fudge was a favourite of those who called during the holidays. When family and friends stopped to wish the family a happy holiday season, a selection of confections and cookies were served with coffee and tea. It was always very exciting for Florence, her brothers and her sisters, because they were also allowed to sample the confections served to the callers.

My grandmother continued the tradition of making candy each holiday season with her daughters, Marie and Glenda. Florence taught them how to make all the family recipes for honey confections, including Nellie's buckwheat honey fudge. My mother remembers spending Saturday afternoons, before Christmas, baking cookies and making candies. Honey candies, placed in little boxes and tied with festive ribbons, made lovely little gifts for teachers and family friends. When my mother was a girl, you could purchase flavour oils from the Watkins Man. Added to the basic candy selection were cool peppermint candies and hot cinnamon candies. Baker's chocolate was also easier to buy and chocolate honey fudge became part of the selection of confections made with honey.

As a small child, I remember there were fantastic selections of candies and cookies, beautifully displayed on serving trays when we went to my grand parents' for Christmas. My mother also made candies when I was a child. I would watch her make fudge, butterscotch and peanut brittle and eagerly ask if I could help. I was so excited when I became old enough to learn to make candy. My mother taught me to make cookies first. I enjoyed rolling out the dough and cutting out the shapes. One Christmas, I decided I would like to make a Christmas gingerbread house. Making the dough and the pieces was no trouble at all, however the instructions called for me to melt sugar to "glue" the pieces of the house together. My mother taught me how to make sugar "glue" and the stages sugar went through when heated. If the sugar got too hot, it would burn. Not heated enough and I could not use the sugar as glue. After making the gingerbread house pieces and "gluing" the pieces together, I had a lovely little gingerbread house. The next step was to decorate the house. I remember getting a piece of cardboard and covering it with foil and using royal icing to cover my gingerbread house with snow. Next, I put tiny, red cinnamon candies and gumdrops on the roof of the house and snow icing around the windows. The only thing missing was a little path to the house. I thought butterscotch candies would make an ideal path. I asked my mother to teach me to make honey butterscotch. Soon my gingerbread house had perfect little butterscotch stepping-stones leading to the door. I was so proud of my gingerbread house with its little butterscotch path. Both recipes are in this book, the gingerbread recipe in Chapter 7 and the butterscotch recipe follows in this chapter.

Ingredients used to Make Confections

Most kitchens have the ingredients used to make candy. To make simple candy, you need sugar and water. You can also use molasses, corn syrup, and honey, in different combinations to give candy a variety of flavours. Corn syrup and honey are also useful in candy making because they prevent formation of large crystals. Dairy products are used to make confections including; butter, milk, and cream, all of which add flavour and a creamy texture to confections. Butter is also excellent at preventing the candy from sticking to the pan and other utensils. To add additional flavour to confections, flavour oils, chocolate, cocoa powder, fruit, nuts and coconut can all be used. Candy can be very plain, such as amber honey candy made with sugar, water and honey, to very fancy squares of honey fudge made with fruits and nuts. No matter which recipe you use, honey confections are sure to be a hit!

Honey in Candy Making

Honey is a wonderful ingredient to add to candy and serves a number of purposes in candy making. Honey can add a variety of flavours, from sweet and delicate with lighter honeys such as clover or oilseed rape, to robust flavours with ivy, heather or buckwheat honeys. Using honey instead of corn syrup adds nutrients and anti-oxidants to the confections that are lacking in candy made with corn syrup. Adding honey also facilitates candy making, by controlling the crystallisation of sugar. Candy can be divided into two general categories; crystalline and non-crystalline candies. Honey fudge is an example of a crystalline candy and consists of small sugar crystals. Non-crystalline candies do not contain these sugar crystals, and includes honey butterscotch or caramels. The first way to keep large sugar crystals from forming in fudge or prevent crystallisation in hard candy is to add fructose or glucose. The presence of fructose and glucose molecules inhibits the sucrose molecules, from forming large crystals, by "getting in the way". Honey is an ideal ingredient for this purpose, as it is primarily fructose and glucose and contains very little sucrose. The second way to prevent sugar crystals from forming, or inhibiting large crystals from forming in candy, is not to stir the pot once it begins to boil. Adding honey to candy ensures fudge with a melt-in-your-mouth texture and hard candies will be crystal free, if you can resist the temptation to stir the pot!

Equipment for Candy Making

Every cook has spoons, spatulas, measuring cups or scales and saucepans. These are the utensils used to make confections. A candy thermometer is not essential however, before you begin to make candy, I recommend you invest in one. You can use the old-fashioned "cold water test" to determine the temperature, but if you are a beginner, you may not recognise when you have "soft ball stage" or "hard ball stage". Being able to interpret the result of dropping your candy mixture into cold water is something that comes with experience. It is much easier to read a thermometer and much more accurate for beginners than to know the stage of the sugar solution. If you do not have a candy thermometer and would like to try a recipe, you can use the "cold water test" to determine when you have reached the desired sugar stage.

Sugar Stages in Candy Making

Sugar goes through a series of stages when you heat it to make candy. Table 6 shows the temperature of each stage, in both Celsius and Fahrenheit, a brief description of the "cold water test" results and the type of candy made at the given stages. Candy making begins when the ingredients are mixed. Stir over medium-low heat until the sugar mixture begins to boil. When the sugar solution has reached a boil, discontinue stirring. It is difficult for some to resist the temptation to stir the sugar mixture, but it is best to leave the pot to boil without mixing. Stirring the sugar solution can introduce sugar crystals and spoil the candy.

Table 6
Stages in Candy Making

Stage	°C	°F	Cold-Water Test*	Recipes
Thread	110-113	230–235	Forms thin threads	Caramel Sauce
Soft Ball	113-116	235–240	A ball which flattens when removed from water	Fudge
Firm Ball	118-121	245–250	A ball which can be flattened by pressure	Caramel
Hard Ball	121-129	250–265	A sturdy ball, but still pliable	Nougat
Soft Crack	132-143	270–290	Hard, pliable threads	Taffy
Hard Crack	149-154	300-310	Hard, brittle threads	Brittle

*To avoid burns, let the sugar syrup cool before examining the syrup dropped into cold water

As the sugar syrup is heated, the first stage in candy making is the "thread stage". Heating to the "thread stage" produces lovely sauces, such as blackberry syrup or caramel sauce. In general, as the temperature increases, the candy becomes harder. When sugar reaches "soft ball stage", the heat is sufficient to make fudge. To make firmer candies, which are still a bit chewy, the sugar solution should be heated to the stages of "firm ball", "hard ball" or "soft crack". These stages produce candies, such as caramels, nougats or taffy. Finally to make hard candies, such as butterscotch or brittles, heat the sugar syrup to temperatures in the "hard crack stage". The temperature of the sugar mixture should not be allowed to go above 170° C or 338° F. At this temperature, the sugar will start to burn and the candy will have a bitter taste.

As you become a more experienced candy maker, you will develop a good feel for when the sugar mixture has reached each stage. I feel confident in my candy making but still rely on my candy thermometer to confirm when I have reached each stage in the candy making process.

Blackberry Honey Syrup

Ingredients	U.S.	Imperial	Metric
Honey	1 C	8 fl oz	240 ml
Fresh Blackberries	¾ C	3.9 oz	110 g

Preparation:
1. In a saucepan, mix honey and blackberries.
2. Place candy thermometer in the saucepan.
3. Heat over medium-low heat.
4. Continue to heat and stir until the mixture is slowly simmering.
5. Discontinue stirring and heat to 221 °F (105° C).
6. Remove from heat.
7. Transfer to jar or container for storage.

Blackberry honey syrup is so easy to make. It can be used as an everyday spread or as a filling in fancy desserts. Use syrup as a spread for toast at breakfast or in a peanut butter sandwich for lunch. Blackberry honey syrup is also good on pancakes or waffles, for a breakfast treat. To make dessert a little more special, use blackberry honey syrup as a topping on ice cream or use as a filling in sandwich cookies or in layer cakes. Delicious when made with apple blossom or lime honey.

Honey Caramel Sauce

Ingredients	U.S.	Imperial	Metric
Single Cream	I C	8 fl oz	240 ml
Sugar	I C	7.1 oz	200 g
Honey	½ C	4 fl oz	120 ml
Butter	¼ C	2 oz	57 g
Salt	¼ tsp	¼ tsp	1.25 ml
Vanilla	I tsp	I tsp	5 ml

Preparation:
1. In a medium saucepan, mix together cream, sugar, honey, butter and salt.
2. Carefully place candy thermometer in the saucepan.
3. Cook over medium-low heat, stirring regularly, until the mixture begins to boil.
4. Discontinue stirring.
5. Heat to 230°F (110°C).
6. Remove from heat.
7. Add vanilla and stir the mixture.
8. Transfer to glass jar.

Honey caramel sauce is very good as a dip for tart apples. The contrast of the warm, honey caramel sauce is a wonderful complement to slices of cool, crisp apple. Why not try granny smith apples, cut into slices and dipped into the warm caramel sauce? Whenever we have caramel sauce with sour apples, I am reminded of the caramel apples we used to eat at Halloween when I was a little girl. Caramel sauce is also delicious on ice cream. Choose dark honey varieties for more robust flavours, such as ivy or heather honey, and higher anti-oxidant content.

Nellie's Buckwheat Honey Fudge

Ingredients	U.S.	Imperial	Metric
Sugar	2 ½ C	17.6 oz	500 g
Double Cream	¾ C	6.3 fl oz	180 ml
Buckwheat Honey	¼ C	2 fl oz	60 ml
Butter	1 T	1 tbsp	14 g
Vanilla	1 T	1 tbsp	15 ml
Butter	Extra nub of butter to grease the pan		

Preparation:
1. Butter a 9x9 inch pan, or line the pan with greaseproof or waxed paper.
2. In a medium saucepan, mix together sugar, double cream, buckwheat honey, and butter.
3. Carefully put the candy thermometer into the sugar mixture.
4. Stir over medium-low heat until mixture begins to boil.
5. When the syrup starts to boil, stop stirring.
6. Let the syrup cook, until the temperature reaches 240° F (115° C).
7. Remove from heat and let the mixture cool to approximately 110° F (43° C).
8. Add vanilla and stir until the fudge thickens and becomes dull in appearance*.
9. Pour the fudge into buttered pan and leave the fudge to cool to room temperature.
10. Once cool, cut the fudge into squares.

*Stirring the fudge, as it cools, prevents large sugar crystals from forming. Small sugar crystals give fudge a smoother texture. If you do not continue to stir, the fudge may become gritty as larger sugar crystals form. The addition of honey helps to prevent large sugar crystals from forming as the fudge cools.

My great-grandmother, Nellie, made buckwheat fudge each year at Christmas time. It was a favourite of family and friends. Nellie made her buckwheat fudge with George's buckwheat honey produced on their farm. My grandmother told me stories of how she and her sisters loved to help make buckwheat fudge because who ever stirred the fudge got to lick the spoon. I sometimes add a cup of chopped hazelnuts to the buckwheat fudge. When I make this fudge with Meggy Jayne and Jasmine they each get a spoon to lick. You can use any variety of dark honey, if you do not have buckwheat honey.

Chocolate Honey Fudge

Ingredients	U.S.	Imperial	Metric
Unsweetened Chocolate	3 Squares	3 oz	85 g
Sugar	3 C	21.3 oz	600 g
Double Cream	¾ C	6 fl oz	180 ml
Honey	¼ C	2 fl oz	60 ml
Butter	1 T	1 tbsp	14 g
Vanilla	2 tsp	2 tsp	10 ml
Butter	Extra nub of butter to grease the pan		

Method:
1. Butter a 9x9 inch pan or line the pan with greaseproof paper or waxed paper.
2. In a medium saucepan, mix together chocolate, sugar, double cream, honey and butter.
3. Carefully put the candy thermometer into the sugar mixture.
4. Stir over medium-low heat until mixture begins to boil.
5. When the syrup starts to boil, stop stirring.
6. Heat the sugar syrup until the temperature reaches 240° F (115° C).
7. Remove from heat and cool the fudge to approximately 110° F (43° C).
8. Add vanilla and stir until the fudge thickens and becomes dull in appearance (This will take a few minutes).
9. Pour the fudge into buttered pan and leave the fudge to cool to room temperature.
10. Once cool, cut the fudge into squares.

Chocolate honey fudge melts in your mouth with a velvety texture. When buying the chocolate to use in this recipe, try to get unsweetened, Baker's chocolate. If you cannot get unsweetened chocolate, look for a dark chocolate with a high percentage of cocoa solids and as little added sugar as you can find. You can make fudge with milk chocolate but the fudge will be very sweet. I prefer the flavour unsweetened chocolate gives the fudge. Both honey and dark chocolate add anti-oxidants to this recipe. Try different dark honeys, like honeydew or tualang.

We make honey fudge at Christmas time and like to experiment with variations on the recipe. Try adding chopped pecans or other chopped nuts, dried fruit, marshmallows or chocolate chips. Place small squares of honey fudge in a decorative box and wrap the box with a bow for a lovely homemade gift.

Chewy Honey Caramels

Ingredients	U.S.	Imperial	Metric
Double Cream	½ C	4 fl oz	120 ml
Milk	½ C	4 fl oz	120 ml
Sugar	1 C	7.1 oz	200 g
Honey	¾ C	6 fl oz	180 ml
Salt	¼ tsp	¼ tsp	1.25 ml
Double Cream (Additional)	½ C	4 fl oz	120 ml
Vanilla	1 tsp	1 tsp	5 ml
Butter	Nub to grease pan		

Preparation:
1. Generously butter a 9 x 9" pan.
2. In a medium saucepan, mix together double cream, milk, sugar, honey and salt.
3. Carefully put the candy thermometer into the sugar mixture.
4. Cook over medium-low heat, stirring regularly, until the mixture begins to boil.
5. Discontinue stirring.
6. Heat to soft-ball stage 235°F (113° C).
7. Gradually add the additional cream, stirring to mix.
8. Continue to cook until temperature reaches 245°F (119° C).
9. Remove from heat.
10. Add vanilla and stir the mixture.
11. Pour the caramel into the well buttered pan.
12. Allow to cool.
13. Cut into pieces, as the caramel cools

Chewy honey caramels have a golden colour and the flavour of honey and vanilla. Soft, yet chewy, these confections will be a family favourite. Try leaving out the vanilla and using a dark honey, like heather or ivy, for caramels with strong honey flavour and high levels of anti-oxidants.

Pink Lemonade Honey Kisses

Ingredients	U.S.	Imperial	Metric
Sugar	2 C	14.2 oz	400 g
Cornstarch (corn flour)	2 T	2 tbsp	18 g
Honey	1 C	8 fl oz	240 ml
Butter	2 T	2 tbsp	28 g
Water	¾ C	6 fl oz	180 ml
Salt	½ tsp	½ tsp	2.5 ml
Lemon Flavour	1 tsp	1 tsp	5 ml
Red Food Colouring	1 or 2 drops		
Butter	To grease pan, butter hands and scissors		
Confectioner's sugar	For dusting candy pieces		
Butter	Nub to grease pan		

Preparation:

1. Generously butter a 9" x 13" pan.
2. In a saucepan, mix together sugar and cornstarch.
3. Stir in the honey, water, butter, and salt.
4. Carefully put the candy thermometer into the sugar mixture.
5. Heat over medium-low heat and stir until the sugar and butter dissolve.
6. Continue to heat until the mixture begins to boil.
7. Discontinue stirring.
8. Heat to 255 ° F (124° C).
9. Remove from heat.
10. Add a few drops of red food colouring and lemon flavour and stir gently.
11. Pour the mixture into a buttered pan.
12. When the candy is cool enough to handle, butter your hands and pull until the candy is light in colour.
13. Roll the candy into a long cable of approximately 1/2 inch diameter.
14. Cut the pink lemonade kisses into 1" pieces with buttered scissors.
15. Dust the candy pieces with powdered sugar.

A kiss is softer than a taffy and should be firm enough to hold its shape. Honey kisses are chewy and have a velvety texture. We make our kisses with apple blossom or lime honey. Meggy Jayne and Jasmine love pink lemonade honey kisses. They are lovely pale pink and have a sweet, lemony flavour. Store in an air-tight container.

Honey Cashew Brittle

Ingredients	U.S.	Imperial	Metric
Sugar	1 C	7 oz	200 g
Honey	½ C	4 fl oz	120 ml
Water	¼ C	2 fl oz	60 ml
Butter	2 T	2 tbsp	28 g
Baking Soda	1 tsp	1 tsp	5 ml
Cashew Nuts, broken	1 ½ C	6.6 oz	188 g
Butter	Nub to grease pan		

Preparation:
1. Generously butter a 9 x 13" pan.
2. In a medium saucepan, mix together sugar, honey and water.
3. Carefully put the candy thermometer into the sugar mixture.
4. Cook on medium-low heat, until the mixture begins to boil, stirring regularly.
5. Discontinue stirring.
6. Heat until the syrup reaches 300°F (149° C).
7. Remove from heat.
8. Add butter, soda and broken cashews and stir the mixture.
9. Pour the frothy sugar mixture onto the well buttered pan.
10. Allow to cool.
11. Remove from pan and break into pieces.

Using Cashew nuts instead of peanuts is an appealing variation on classic peanut brittle. The buttery flavour of cashew nuts mingled with the floral taste of honey adds a bit of sophistication to nut brittle. Try using different combinations of nuts and honey varieties to vary the taste. For traditional peanut brittle, use raw peanuts.

Honey Butterscotch

Ingredients	U.S.	Imperial	Metric
Brown Sugar, Packed	1 C	7.7 oz	217 g
Honey	2 T	2 tbsp	30 ml
Water	½ C	4 fl oz	120 ml
Salt	¼ tsp	¼ tsp	1.25 ml
Butter	3 T	3 tbsp	45 ml
Lemon Extract	2 drops	2 drops	2 drops
Butter	Nub to grease pan		

Preparation:
1. Generously butter a 9 x 13" pan.
1. Mix together brown sugar, honey, salt and water, in a medium saucepan.
2. Carefully put the candy thermometer into the sugar mixture.
3. Cook over medium-low heat, stirring regularly, until the mixture begins to boil.
4. Discontinue stirring.
5. Heat until the syrup reaches 250°F (121° C).
6. Add butter and cook to 300°F (149° C).
7. Remove from heat.
8. Add lemon extract and stir the mixture.
9. Pour into buttered pan*.
10. While still warm, score the butterscotch with a knife.
11. Allow to cool completely.
12. Remove from pan and break into pieces along the scored lines.

*If you wish, you may pour the butterscotch in circles on a buttered tray.

Honey butterscotch is the first candy my mother taught me to make. The Christmas I made my first gingerbread house, I made my first butterscotch candies for the tiny path that led to my gingerbread house's door. I poured the rest of the recipe into a buttered baking pan.

Amber Honey Candy

Ingredients	U.S.	Imperial	Metric
Sugar	1 ¾ C	6.1 oz	350 g
Honey	½ C	4 fl oz	120 ml
Water	⅓ C	2.7 fl oz	80 ml
Butter	Nub to grease pan		

Preparation:
1. Butter a 9" x 9" pan.
2. In a medium saucepan, mix together sugar, honey, and water.
3. Position the candy thermometer in the saucepan.
4. Heat over medium-low, stirring regularly, until the mixture begins to boil.
5. Discontinue stirring.
6. Heat to 310°F (154° C).
7. Remove from heat.
8. Pour liquid candy mixture into the buttered pan.

Pieces of amber honey candy look like glistening gems. The colour can range from pale topaz to rich amber brown depending on the variety of honey you choose. Why not experiment with different varieties of honey. Try lime, heather, ivy or buckwheat honey for a range of golden colours. You can also add colour and flavour oils to this basic honey candy recipe. Variations include peppermint oil and blue food colouring, wintergreen oil and green food colouring or lemon flavouring with yellow food colouring.

When you crave something sweet, choose honey confections. You will benefit from higher levels of anti-oxidants than if you have candies made with sugar and corn syrup.

Chapter 9

Aunt Bea's Remedies

At the turn of the 19th century, going to the doctor was an expensive luxury. Every home had a book of practical recipes. Contents included a medical section containing formulae for making home remedies for common ailments. When my grandmother was a girl, her Aunt Beatrice was charged with preparing the family's home remedies. Aunt Bea knew how to make all the remedies needed to treat minor ailments, using bee products, her larder and her garden as her pharmacy. Aunt Bea could mix together a remedy for nearly any minor complaint. Local plants were dried and stored in jars for Aunt Bea to use in her concoctions. Many of the formulae contained honey, propolis and beeswax.

When I was five years old my family visited my grandmother, grandfather and Aunt Bea, in the summertime. I do not have many memories of that summer, but do remember one event very vividly. My brother, David and I went exploring in the back garden. The kitchen door led to a patio with a slate wall. Above the wall, laid a hill covered with rose bushes, beyond which were the rhubarb and strawberry patches. The strawberry patch was one of our favourite places where we enjoyed sampling the sweet, ripe strawberries. One sunny afternoon, David and I decided to go up to the strawberry patch. We walked past the patio and the long way around, on the footpath that led up the hill. The path wound slowly up the hilly garden, to the strawberries and was much less steep than going up and over the slate wall. We spent some time at the strawberry patch and then decided to go back down the hill to the kitchen. It seemed like a good idea to me to go a more direct route to the house, rather than going the long way around. I started down the steep hill. I lost my footing on the hill and tumbled down, through the rose bushes, over the slate wall, landing flat on the patio floor. My trip down the hill and through the rose bushes left me scratched and bruised. David tumbled down the hill behind me. Somehow my brother landed unharmed. I remember thinking it was very unfair that I was covered with scratches and David tumbled down without a scratch! Aunt Bea took care of me after my tumble. She changed me into one of my grandfather's tee shirts and after cleaning my scratches, applied beeswax cerate. I looked a bit scruffy in my grandfather's large tee shirt and dotted with cerate but the cerate did the trick. It took the redness out of the scratches and they healed in no time at all.

Aunt Bea's remedies work because of the medicinal properties of the ingredients, such as beeswax cerate to heal cuts and burns. Now it is known that honey prevents infections and the combination of beeswax, olive oil and honey speeds wound healing. Aunt Bea's remedies are useful for colds, sore throats, hay fever, as well as skin problems, such as eczema and psoriasis. Before using Aunt Beas remedies, check with you doctor and seek further medical advice if symptoms persist.

Honey Energy Drink

When my great-grandfather was a farmer, after a long day's work, he would enjoy a homemade energy drink called switchel. Workers needed something to replenish the minerals and fluids lost during their physical labour. Many of these energy drinks contained honey.

Haymaker's Switchel

Ingredients	U.S.	Imperial	Metric
Water	8 C	64 fl oz	1.9 L
Honey	1 C	8 fl oz	240 ml
Brown Sugar, Packed	1 C	7.7 oz	217 g
Vinegar*	1 C	8 fl oz	240 ml
Ground Cinnamon	1 tsp	1 tsp	5 ml
Ground Ginger	2 tsp	2 tsp	10 ml

*Although switchel contains a large amount of vinegar, the taste is very pleasant.

Preparation:
1. Mix all ingredients in a large bowl.
2. Transfer to a bottle with lid.
3. Store in a cool place.
4. Serve neat or with ice.

George often enjoyed a cool glass of refreshing "ginger water" or switchel when he returned from the fields. One day George found a rattlesnake's nest in his hay field and he said he really needed a glass of switchel after dealing with those snakes. The glucose, fructose and sucrose gave the workers energy and the honey provided minerals, which were lost in sweat during exercise. Haymaker's switchel is so easy to make. You just measure, mix and cool for a refreshing energy drink. Aunt Bea and Nellie used apple cider vinegar and buckwheat honey in their switchel. You can try different combinations of honey and vinegar types for varied flavours. Raspberry vinegar and blossom honey have a nice fruity flavour. We like to use apple cider vinegar and buckwheat honey, my great grandmother and Aunt Bea's recipe.

Honey for Colds and Flu

When someone showed the first signs of a cold or flu developing, Aunt Bea would prescribe a spoonful of oxymel. Oxymel is a blend of honey and vinegar, which helps alleviate some of the symptoms of colds and flu and helps speed recovery. There is now scientific evidence that oxymel is therapeutic to treat colds. The anti-viral activity in honey may decrease the duration of the illness and honey certainly soothes a sore throat. In addition, the anti-oxidants and vitamin C in honey may help speed recovery. Aunt Bea would have used my great-grandfather's buckwheat honey in her oxymel, however any honey will work a treat!

Aunt Bea's Oxymel

Ingredients	U.S.	Imperial	Metric
Vinegar	¼ C	2 fl oz	60 ml
Buckwheat Honey	I C	8 fl oz	240 ml

Preparation:
1. Stir to mix vinegar into a glass of warm water or tea.
2. Store in a jar.

Oxymel is acidulous syrup, made of honey and vinegar and used as a remedy for sore throats and coughs. Oxymel is very easy to make, using only two ingredients, honey and vinegar. It is possible to change the character of the oxymel by varying the type of honey and vinegar used. The secret to making a good oxymel is getting the consistency correct. Use dark honey for added anti-oxidants and nutrients. For a sore throat or cough take a spoonful as necessary. For Jasmine and Meggy Jayne, I use blossom honey and apple cider vinegar. Aunt Bea's Oxymel soothes their sore throats and lifts their spirits. A spoonful of honey is always good medicine. In addition to medicinal use, oxymel is also a dietary supplement and aids to digestion. Take one teaspoon a day as a dietary supplement.

Oxymel and Propolis Gargle

Ingredients	U.S.	Imperial	Metric
Aunt Bea's Oxymel	I tsp	I tsp	5 ml
Warm Tea or Water	¼ C	2 fl oz	60 ml
Propolis Tincture	a few drops		

Preparation:
1. Stir a teaspoon of oxymel into a glass for warm water.
2. Add a few drops of propolis tincture.
3. Stir to mix thoroughly.
4. Take a sip and gargle to alleviate sore throat.
5. Do not swallow, but discard after gargling.

While neat oxymel soothes the throat, an oxymel gargle will help further ease throat pain and acts as an anti-septic to treat the throat infection. Propolis tincture adds anti-viral activity to the oxymel gargle. Use oxymel gargle four to six times a day. Make fresh for each gargle.

Propolis Cough Drops

Ingredients	U.S.	Imperial	Metric
Sugar	1 ¾ C	12.3 oz	350 g
Honey	½ C	4 fl oz	120 ml
Water	⅓ C	2.7 fl oz	80 ml
Propolis Tincture	A few drops		
Butter	Nub for buttering pan		

Preparation:
1. Butter a baking tray.
2. In a medium saucepan, mix together sugar, honey, and water.
3. Position the candy thermometer in the saucepan.
4. Cook over medium-low heat, stirring regularly, until the mixture begins to boil.
5. Discontinue stirring.
6. Heat to 310°F (154° C).
7. Remove from heat.
8. Add the propolis tincture and stir.
9. Pour liquid into the buttered pan.
10. When the candy is cool enough to handle, stretch the candy and twist into a half inch diameter rope.
11. Using buttered scissors, cut the cough candy into one inch lengths and place in powdered sugar.

The anti-viral activity of propolis can help you fight a developing cold. Propolis cough drops can be taken to sooth a sore throat, help treat a cold and to treat a cough. The propolis gives the drops a tingly feeling, suggesting the propolis is fighting the virus. Honey also soothes and makes the drops efficacious. Store propolis cough drops in an airtight container.

Honey for Hay Fever

Many of us suffer with hay fever. Hay fever is a pollen allergy and is associated with itchy eyes, sneezing, coughing, and wheezing. Honey can serve as a simple remedy to help minimise the symptoms of hay fever. Honey is made from the nectar of various flowers and also contains pollen from the flowers the bees visit while collecting nectar. Because honey contains pollen, honey can be used to de-sensitise allergy sufferers to pollen. By taking a spoonful of honey each day, beginning a few months before hay fever season, hay fever sufferers can lessen the degree of allergic reaction to pollen.

If you are lucky enough to know what pollen you are allergic too, you can choose a honey that is made with the nectar from this plant. For instance, if you are allergic to tree pollens, choose a tree honey. A second recommendation is to get honey made in the season you suffer hay fever. So if you get hay fever in the autumn, choose an autumn honey, because it will contain pollen from the season in which the plants you are allergic to flower. If you do not know which pollen is affecting you, then selecting a honey from a "local" beekeeper will be of value because this honey will contain local pollens. Finally, a brief note about local honey. Local honey does not necessarily need to be honey from a beekeeper in your community. As long as the honey is from a region with the same plants, the honey will work. The most important thing is the pollen in the honey.

Honey for Indigestion

Aunt Bea treated indigestion or dyspepsia with a tonic made from chamomile flowers and honey. Today we can easily source chamomile tea for the same purpose.

Honey & Chamomile Indigestion Tonic

Ingredients	U.S.	Imperial	Metric
Chamomile Tea		1 bag	
Boiling Water	1 C	8 fl oz	240 ml
Honey	2 tsp	2 tsp	10 ml
Fresh Ginger (grated)	½ tsp	½ tsp	2.5 ml

Preparation
1. Add tea bag and grated ginger to cup. Add boiling water to prepare the chamomile tea.
2. Remove tea bag and stir in honey. Stir until the honey dissolves.

Aunt Bea recommended a cupful of honey and chamomile tonic in the morning to strengthen the digestive organs, and a teacupful for aged persons a couple of hours before dinner. Give tonic to fretful children in small doses to improve their appetites. Ginger is an anti-emetic and this coupled with the soothing properties of honey and chamomile, gives this remedy its efficacy. Aunt Bea would also make honey and chamomile tonic for women "in a family way". The ginger would help relieve morning sickness.

Skin Conditions

Honey has anti-bacterial qualities, which contribute to the health benefits associated with honey in the skin preperations. The anti-bacterial actions of honey are a result of five mechanisms, four of which are present in all honey varieties. The four actions present in all honey include: low pH or acidity, high osmolarity, production of peroxide, and the presence of the enzyme lysozyme. The low pH of honey kills bacteria through the action of exposure to acidic environments. High osmolarity draws water out of the bacteria, leading to bacterial death. Peroxide kills bacteria through oxidative attack of bacteria. Hydrogen peroxide is made by the enzyme glucose oxidase in honey. Hydrogen peroxide reacts with iron to produce free radicals, which kill bacteria. Finally, the enzyme lysozyme kills bacteria by breaking down the bacterial cell wall.

In addition to the four anti-bacterial mechanisms seen in all honey, some honey also contains a secondary anti-bacterial activity, which relates to the botanical source of the honey. This secondary anti-bacterial activity, is well known in manuka honey. Manuka honey contains methylglyoxal, which is toxic to bacteria. Because the first four anti-bacterial mechanisms are present in all honey, any honey variety applied to minor cuts and scrapes can help prevent infections.

Honey and beeswax are very useful ingredients in remedies to treat skin conditions. Additional ingredients are sometimes added to these preparations, but many conditions can be helped or even cured by simple combinations of oil and honey or blends of beeswax, oil and honey. Simple ointments help alleviate eczema and psoriasis and chapped skin is well served by salves and burns heal more quickly when a simple beeswax cerate is applied.

There are two main categories of preparations for oil-based skin remedies, ointments, (sometimes called salves) and cerates. Both ointments and cerates contain no water and protect, heal and soothe skin conditions. Ointments often have the consistency of butter while cerates tend to be much firmer because of the higher amount of wax they contain. Ointments and cerates, are made with a number of active ingredients, including honey, beeswax and propolis. Preperations made with beeswax and honey sooth skin and facilitate healing.

Blush Lip Balm

Ingredients	Imperial	Metric
Beeswax	0.7 oz	20 g
Sweet Almond Oil*	2.1 oz	60 g
Cocoa Butter	0.7 oz	20 g
Alkanet Root	Small Spoonful	
Cotton Fabric	4" x 4" Piece	
Twine	Approximately 6 inch piece	

*Use rapeseed or sunflower oil instead of almond oil for those with nut allergies.

Preparation:
1. Weigh beeswax, cocoa butter and sweet almond oil into the bowl or Pyrex jug.
2. Heat the ingredients in the microwave or bain-marie until the beeswax and cocoa butter have melted.
3. Place a small spoonful of alkanet in the centre of the fabric square. Bunch the fabric and tie the alkanet root into the cotton fabric with twine.
4. Drop the alkanet sachet into the warm wax and oil mixture.
5. Soak the alkanet in the wax mixture until the desired colour is reached.
6. Remove alkanet sachet.
7. Pour into small pots.

Blush lip balm will prevent redness and heal dry chapped lips. Alkanet root is high in anti-oxidants and will add anti-oxidants, as well as colour, to this lip balm. For a light blush-coloured lip balm, only leave the alkanet sachet in the warm balm for a short time. Deeper pinks and reds can be achieved by leaving the sachet of alkanet in the liquid longer. You can leave the alkanet sachet in the lip balm until it has cooled and then warm the balm again to remove the sachet and pour into pots. The darker the colour the higher the anti-oxidant levels in the lip balm.

Propolis Ointment

Ingredients	Imperial	Metric
Beeswax	0.5 oz	15 g
Sweet Almond Oil	1.2 oz	35 g
Propolis Tincture*	A Few Drops	

*Propolis tinctures made with polypropylene glycol will be easier to mix into ointment than alcohol tinctures.

Preparation:
1. Weigh beeswax and sweet almond oil into a glass bowl.
2. Transfer bowl to microwave or bain-marie.
3. Heat until beeswax has melted, stirring often.
4. Remove from heat, add propolis and stir.
5. Pour into glass jar while still hot.

Propolis ointment is a very basic ointment for use on dry skin, on chapped lips and cold sores. The anti-viral activity of propolis will speed healing of the cold sore. Many old remedies contained turpentine, which is prepared by distilling the aromatic resins from pine trees. Today we would not use turpentine. Propolis is also made from tree resin has similar characteristics. Propolis is anti-bacterial, anti-seborrheic and has other healing properties of aromatic tree resins.

Honey Cerate for Burns and Cuts

Ingredients	Imperial	Metric
Beeswax	0.7 oz	20 g
Olive Oil	1.4 oz	40 g
Buckwheat Honey	1.4 oz	40 g

Preparation:
1. Warm beeswax and olive oil in bain-marie or microwave until beeswax has melted.
2. Remove from heat.
3. Slowly add honey to the beeswax and olive oil mixture, stirring continuously.
4. Continue to stir until completely cool to prevent the honey from separating from the oil and beeswax.

The simple combination of olive oil, honey and beeswax provides powerful healing properties to this cerate. There have been studies showing the use of a combination of olive oil, honey and beeswax significantly speeds wound healing. We use this simple cerate anytime we have a cut, blister or burn. Indeed, I once picked up a candle, which was recently, blown out and spilled hot wax down my arm. I instinctively grabbed the cerate and applied it to my skin after peeling off the wax. I was sure this would minimise the burn caused by the hot wax. What I did not expect was that the cerate would completely prevent any redness from appearing or blisters from forming. I am aware that this is anecdotal evidence, for the efficacy of this remedy, but this is further support of the scientific research, which demonstrates this combination of ingredients can facilitate wound healing.

Waxeline

Waxeline is an all-natural product without any petroleum-based ingredients.

Ingredients	Imperial	Metric
Beeswax	0.7 oz	20 g
Rapeseed Oil	2.8 oz	80 g

Preparation:
1. Place rapeseed oil and beeswax in a Pyrex jug.
2. Warm beeswax and rapeseed oil in bain-marie or microwave until beeswax has melted.
3. Remove from heat.
4. Stir and pour into pots.

Waxeline is a bee-based version of Vaseline. Vaseline was introduced, when Aunt Bea was a young woman and received a lot of publicity because it was taken on polar expeditions. Aunt Bea thought she could make something just as nice with beeswax. Waxeline is used the same way you use Vaseline, to moisturise skin and adds a protective barrier.

Sunburn Remedy

Ingredients	U.S.	Imperial	Metric
Vinegar	½ C	4 fl oz	120 ml
Honey	2 T	2 tbsp	30 ml

Preparation:
1. Mix the honey into the vinegar and stir well, to thoroughly dissolve the honey.
2. Use a cotton or cloth to apply gently to the affected area.

Today we understand the risks associated with sunburn and understand the best remedy for sunburn is prevention. Always apply sunscreen and limit exposure to the sun. Sometimes, there are occasions when we misjudge the length of time we are in the sun and are exposed to more sun than we intended. In these situations, Aunt Bea's sunburn Remedy can sooth the skin and facilitate healing.

Insomnia

Aunt Bea had a few magical remedies for insomnia, which worked very well on children; Sleepeasy Potion, Salve and Spell. Children seem to think that all the fun begins after they go to bed. Adults get out the treats and games as soon as they are asleep. No one, especially children want to miss a chance at fun. Like the generations before them, Meggy Jayne and Jasmine sometimes do not want to go to bed. You can coax children to sleep with a warm drink, a story or a little bit of magic. Aunt Bea combined all of these for perfect remedies for children reluctant to go to bed.

Sleepeasy Potion

Ingredients	U.S.	Imperial	Metric
Milk	I C	8 fl oz	240 ml
Honey	I tsp	I tsp	5 ml
Grated Nutmeg	Sprinkle		

Preparation:
1. Warm the milk in a pan or microwave. Take care to keep the milk from becoming too hot for children.
2. Add honey and stir until honey, is completely dissolved.
3. Sprinkle a little magic nutmeg on the top.

Take Sleepeasy Potion before bed will facilitate sleep in those reluctant little ones. When combined with the Sleepeasy Spell, a good night's sleep is sure to be. A little magic will help any child get to sleep. The Sleepeasy Spell is a magical way Aunt Bea used to put small children to bed.

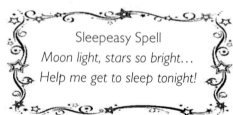

Sleepeasy Spell
Moon light, stars so bright…
Help me get to sleep tonight!

Simply help the child say the Sleepeasy Spell after drinking Sleepeasy Potion or after applying Sleepeasy Salve. Guaranteed to lead to sweet dreams of wildflowers and honey bees.

Sleepeasy Salve

Ingredients	Imperial	Metric
Beeswax	0.7 oz	20 g
Rape Seed Oil	I oz	30 g
Lavender Essential Oil	1-2 Drops	

Preparation:
1. Put beeswax and rape seed oil in a Pyrex jug.
2. Warm beeswax and rapeseed oil in bain-marie or microwave until beeswax has melted.
3. Remove from heat.
4. Add lavender essential oil and stir.
5. Pour salve into a pretty, little jar or pot while still liquid.

Children can apply Sleepeasy Salve before they go to bed each night. Dab a little on the chest or behind their ears. The gentle aroma of lavender and honey-scented beeswax is sure to get them off to sleep.

Chapter 10

Anti-oxidant Soaps

Anti-oxidants used in skin care products can protect skin from oxidative stress and help slow the signs of aging. Increasing the anti-oxidant levels in our skin can help combat the oxidants we are expose to in daily life. Every day, we are exposed to UV light, chemicals and other pro-oxidants. The polyphenolic compounds in honey, beeswax and propolis, when applied to the skin, will increase the levels of anti-oxidants in skin, decrease the signs of aging and protect skin from oxidative stress. Using beauty products made with bee products, can keep you looking young and beautiful.

For many, their daily beauty routine begins with washing their face. Honey soap gently cleanses skin, provides anti-oxidants and, because of honey's emollient properties, leaves skin well moisturised. Making your own honey soap is not difficult using my method. You can make the following recipes for cold-processed soap with minimal equipment and use the soap the same day you make it. For more details about this method and soap making, see *Dr Sara's Honey Potions*.

You make soap by reacting oils and fats with an aqueous hydroxide solution, commonly called lye. Saponification is the chemical reaction, of oils with hydroxide, to produce soap and glycerine. This method uses a very concentrated lye solution. Because the solution is very concentrated, the saponification reaction completes extremely quickly. In order to formulate a soap recipe, the amount of sodium hydroxide needed to react, with each specific oil is calculated, using the saponification values. The following recipes have been formulated using the saponification value for each oil in the recipe you may not substitute other oils for those in the recipe.

Equipment for Soapmaking

Most of the equipment needed to make your own soap you will have in your home. There are two items which are essential for soap making. The first is a weighing scale. It is advisable that the scale measures in increments of one gram for making cold-processed soap. The second piece of equipment needed for soap making is a hand blender or stick mixer. You will also need to melt the oils in either in a microwave or on the stovetop. Either will work and does not alter the quality of the soap. Other things you will need are glass or Pyrex jars, measuring cups and a pan or plastic container. Measuring spoons, wooden spoon, plastic film, gloves and towels are also useful when making your own soap.

Soap Making Ingredients

You can buy most of the ingredients used in soap making in local shops. To make soap, you need three ingredients; water, sodium hydroxide and oil. The water is used to dissolve the sodium hydroxide. You may want to use bottled water in your soap. The second ingredient you need is sodium hydroxide. You can buy this in a number of shops. Sodium hydroxide, is frequently called caustic soda. When you buy sodium hydroxide, look for a purity of at least 98%. If the caustic soda is less pure, it will not work well in soap making recipes. The last ingredient you need is oil. You can use a variety of oils to make soap as well as fats. Today supermarkets and ethnic shops stock a wide variety of oils. If you are looking for specialty oils you can get oils for soap making online from soap making supply companies. Other ingredients you may want to add to your soap including honey, colour and fragrance.

Making Soap

When making soap the first thing you do is to make the lye solution. Pour the sodium hydroxide (NaOH) crystals into the water and stir until the hydroxide is suspended. Put the lye solution to the side and get the oils ready. Measure the oils by weight. Slowly pour the lye solution into the bowl containing the oils while stirring with a stick blender with the power off. After adding all the lye solution, begin mixing with the power on. Continue to stir until the soap mixture begins to thicken. The soap will begin at a consistency like a thin sauce and continue to thicken until it is like custard and will then resemble a thick batter. Pour the soap into a pan and cover the pan with cling film and towels to keep the soap from loosing heat. The soap will continue to warm until it heats up from the centre to the edges. This is the saponification reaction converting the oil into soap. Once the reaction has reached the edge of the pan, unwrap the soap and allow it to cool.

Buckwheat Honey & Molasses Anti-oxidant Soap

This is a cold-processed recipe for soap high in polyphenol anti-oxidants. Buckwheat honey and molasses provide anti-oxidant activity including free radical scavenging, iron chelation and prevention of the generation of free radicals, all of which will help prevent the signs of aging.

Ingredients	Imperial	Metric
Sodium Hydroxide (NaOH)	4.6 oz	131 g
Water	8.2 oz	233 g
Olive oil	5.3 oz	150 g
Sunflower Oil	7.9 oz	225 g
Coconut oil	18.5 oz	525 g
Cocoa Butter	0.9 oz	25 g
Buckwheat Honey	3 tbsp	45 ml

Molasses	3 tbsp	45 ml
Coconut Oil	Small nub for greasing pan	

Making the Lye Solution
1. Put on your gloves before you begin.
2. Place the large glass jar or Pyrex jug on the scale and weigh the water into the jar.
3. Place the small plastic or glass bowl on the scale, and carefully weigh the sodium hydroxide into the bowl.
4. Pour the sodium hydroxide into the water and mix with a wooden spoon until the crystals are suspended in the water (The solution will warm up as the sodium hydroxide dissolves into the water).*

*Make sure the lye solution it is out of reach of children & pets. Do not inhale fumes.

Preparing the Oils
1. Place the glass bowl on the scale and add the coconut oil and cocoa butter.
2. Transfer the glass bowl with the solid coconut oil and cocoa butter to the microwave or bain-marie and heat just until the oils have melted.
3. Weigh the olive oil and sunflower oil into a large glass bowl.
4. Add the melted cocoa butter and coconut oil to the larger glass bowl containing the sunflower and olive oil.
5. Add the heather honey and molasses to the oil.
6. Stir the oils, honey an molasses together using the hand blender.

Mixing the Soap
1. Slowly pour the lye solution into the large bowl containing the oil mixture, stirring, slowly, with the stick mixer off.
2. After all the lye has been added and mixed, with the oils, begin to stir the soap with the stick mixer powered on.
3. The soap mixture will become more opaque and begin to thicken.
4. Stir until the soap mixture is like a cake batter.
5. Pour the soap mixture into the mould or pan.
6. Wrap the soap with cling film and cover the soap with layers of towels to keep it from loosing heat.

Buckwheat honey and molasses soap has a subtle caramel smell with hints of liquorish. As you use the soap, you will be protecting and nourishing your skin. If you do not have buckwheat honey, you can use any dark honey in this recipe. Honey varieties such as forest, honeydew or heather honey will work very well and add anti-oxidants.

Witch Hazel & Propolis Anti-bacterial Soap

For those with problem skin, honey, witch hazel and propolis soap is anti-biotic and may help.

Ingredients	Imperial	Metric
Sodium Hydroxide (NaOH)	4.6 oz	130 g
Water	8.3 oz	235 g
Olive oil	14.6 oz	415 g
Sunflower Oil	7.9 oz	225 g
Coconut oil	18.5 oz	525 g
Witch Hazel	1 tbsp	15 ml
Heather Honey	3 tbsp	45 ml
Propolis Tincture	5 Drops	
Coconut Oil	Small nub for greasing pan	

Making the Lye Solution

1. Put on your gloves before you begin.
2. Place the large glass jar or Pyrex jug on the scale and weigh the water into the jar.
3. Place the small plastic or glass bowl on the scale, and carefully weigh the sodium hydroxide into the bowl.
4. Add the sodium hydroxide to the water and mix with a wooden spoon until the crystals are suspended in the water (The solution will warm up as the sodium hydroxide dissolves into the water).*

*Make sure the lye solution it is out of reach of children & pets. Take care not to inhale fumes, as the caustic soda dissolves.

Measuring the Oils

1. Place a large glass bowl on the scale and add the coconut oil.
2. Transfer the glass bowl with the solid coconut oil to the microwave or bain-marie and heat just until the oil has melted.
3. Weigh the olive and sunflower oil into the glass bowl containing the melted coconut oil.
4. Add the witch hazel, heather honey and propolis to the oil and stir using the hand blender.

Mixing the Soap

1. Pour the lye solution into oil mixture. Stir slowly with the stick mixture off.
2. After you have added all the lye, start mixing with the stick mixer powered on.
3. The soap mixture will become more opaque and begin to thicken.
4. Stir until the soap mixture is like a cake batter.
5. Pour the soap mixture into the pan.
6. Wrap the soap with cling film and cover the soap with layers of towels to keep the soap from loosing heat.
7. When the soap has completed processing, unwrap the soap and cut the soap into pieces.

Witch hazel and propolis soap has a dark brown colour and a slight smell of propolis. Using propolis soap on a regular basis will remove accumulated sebum. The anti-bacterial and anti-seborrheic activity of propolis can help prevent acne. Witch hazel is anti-inflammatory and can reduce redness and swelling, while the anti-oxidants in this soap will also help reduce signs of aging.

Honey Savonnettes

Savonnettes are little soaps made with bar soap and other ingredients including herbs and honey. These little soap balls were popular in the 1800s and were made in many homes, including my great-grandparent's. To make savonnettes, bar soap is grated and mixed with other ingredients to add colour and fragrance. The ingredients are "crutched" or held together with a binding ingredient. One of the ingredients used to crutch the soap was honey. Honey makes the grated soap, sticky so it can easily be shaped. Savonnettes are often decorated with dried flowers or herbs. Honey savonnettes adorned and perfumed the washing chambers of many homes when special guests came to visit.

Making savonnettes is so easy. There is no heat involved and you can make such a variety by changing the ingredients you add to the mix. Jasmine and Meghan enjoy making savonnettes. They like using their hands to mix the soap with the honey and choosing their own combination of ingredients to add. Honey adds emollient qualities and anti-oxidants to the savonnettes. My great-grandmother and her daughters often made savonnettes with their family honey.

To make savonnettes, simply grate solid soap. I use my honey soaps, but a purchased bar soap would also be suitable. The grated soap is then mixed with just enough honey to make the soap bind together. The soap is formed into various shapes and flowers and other botanicals pushed into the top while they are still wet. Savonnettes were used as decorative toilet soaps and often had roses, lavender and spices pushed into them. If you would like to add fragrance to your savonnettes, you may wish to start with an unscented bar of soap. The advantage of using an unscented, honey-free soap as a base is that the starting soap is unscented and white in colour. This is a blank pallet. Use the unscented soap recipe, which follows in the Glycerine Soap section of this chapter. The other option is to use a base soap, which has colour and fragrance already added. You can still add additional ingredients and shape the savonnettes to make beautiful little toilet soaps. If you use handmade soap, the savonnettes will be very nice, indeed.

Rose Savonnettes with Dried Roses

Grate two bars of rose soap, add approximately two tablespoons honey and a few dried rose petals. Once the honey and soap have been mixed, roll the soap into little ball-shaped savonnettes. Finish by placing a few dried roses in the top of the soap. Allow the savonnettes to air dry for a few days.

Gingerbread Savonnettes

Begin by grating two bars of soap. To make gingerbread savonnettes, buckwheat honey and molasses soap works well. Add a few tablespoons of honey to the grate soap and add a few drops of gingerbread fragrance. Shape into ovals and press a star anise into the top. Gingerbread savonnettes smell wonderful and are very festive with their dark gingerbread colour, fragrance and beautiful star on top.

Green Gardener's Savonnettes

First, grate two bars of white soap, such as the unscented soap in this chapter. Crutch the grated soap with a few tablespoons of ivy honey. Add a small spoonful of green clay to the soap, mix and shape into soap balls. The green clay gives these savonnettes a lovely garden green colour. Why not add a few drops of basil or rosemary essential oil to give the soap balls a fresh, garden fragrance.

Savonnettes with Heather Honey and Propolis

Begin by grating two bars of a lightly scented soap. To the grated soap, add a few tablespoons of heather honey and a few drops of propolis tincture. Heather honey and propolis provide high anti-oxidants and the propolis adds antibiotic activity.

Glycerine Soap

Glycerine soap is frequently associated with "melt and pour" soap because glycerine soap melts at low temperatures (between 50° C and 65° C). You can melt a small piece of glycerine soap, add a little colour, honey and fragrance and pour the soap into shaped moulds. Making soap with glycerine soap base is an activity that children can enjoy with an adult. You can easily make your own glycerine soap at home with cold-processed soap by adding glycerine and water. The method presented here is very easy to do and results in an opaque glycerine soap base that can be use to make a variety of soaps.

Unscented Cold-Processed Soap

Begin by making an unscented soap without colour or honey, using the following recipe. This unscented soap will be made into the glycerine soap base.

Ingredients	Imperial	Metric
Sodium Hydroxide (NaOH)	4.6 oz	130 g
Water	8.3 oz	235 g
Olive oil	14.6 oz	415 g
Rape Seed Oil	7.9 oz	225 g
Coconut Oil	18.5 oz	525 g

Making the Lye Solution

1. Put on your gloves before you begin.
2. Weigh the water into the jar or Pyrex jug on the scale.
3. Place the small plastic or glass bowl on the scale and weigh the sodium hydroxide into the bowl.
4. Slowly pour the sodium hydroxide into the water and mix with a wooden spoon until the hydroxide is suspended in water (The solution will warm up as the sodium hydroxide dissolves into the water).
5. Put the lye solution to the side.

Preparing the Oils

1. Place the glass bowl on the scale and add the coconut oil.
2. Transfer the solid coconut oil to the microwave or bain-marie and heat just until melted.
3. Weigh the olive and rapeseed oil into the glass bowl containing the melted coconut oil and stir.

Mixing the Soap

1. Pour the lye solution into the bowl containing the oils and stir with the mixer off.
2. After you add the lye, stir with the stick mixer powered on.
3. The soap mixture will become more opaque and begin to thicken.
4. Continue to stir until the soap mixture is thick like cake batter.
5. Pour the soap mixture into the pan
6. Wrap the soap with cling film and cover the soap with layers of towels to keep the soap from loosing heat.

After the soap has finished processing, allow the soap to cool. The unscented soap can be stored and used as needed. I keep a stock of unscented soap around to make up glycerine soap base as needed.

Making the Glycerine Soap Base

To make the soap base, you will need glycerine. Glycerine is a by-product of commercial soap manufacturing, made from animal fats and vegetable oil. Vegetable derived glycerine can be purchase if you prefer.

Ingredients	Imperial	Metric
Unscented Soap	3.5 oz	100 g
Glycerine	3.5 oz	80 g
Water	1.4 oz	20 g

Method
1. Grate the unscented soap and place in a glass bowl
2. Add glycerine and water
3. Stir the soap, glycerine and water
4. Heat on bain-marie stirring often
5. When the soap has dissolved into the glycerine, pour the soap into a pan or plastic container and let cool

Use the glycerine soap base to make soaps in different colours, with different fragrances and honey. Melt as much or as little as you like and add anti-oxidant ingredients, such as dark honey or vitamin E. You can add any honey, colour and fragrance to the soap base. Another advantage to glycerine soap is that you can make a small quantity of each kind, allowing you to have a variety of soaps to use rather than being restricted to one type of soap from a large batch. Store glycerine soap in an air-tight container.

Chapter 11

Age Defying Moisture Creams

Using a moisture cream containing anti-oxidants can delay the signs of aging and even make you look younger. There are numerous brands boasting anti-aging effects with added anti-oxidants such as vitamin E, coenzyme Q, and plant extracts containing polyphenols. All of these anti-oxidants are free radical scavengers and inhibit oxidative damage to skin. The price tags on brand name anti-aging creams can be very high. Making your own anti-aging moisture cream does not need to be too complicated and certainly is less expensive than name brand creams. Another advantage to making your own moisturiser is that you can choose the ingredients you would like to use and formulate a cream for your own skin type and needs.

Equipment for Making Moisture Cream

The equipment needed, to make your own anti-oxidant creams, is very similar to the equipment needed to make your own soap. Many of the items you will have in your home. Just as with making soap, the two vital pieces of equipment are a scale and a mixer. You will measure the ingredients by weight, rather than volume for most ingredients in the cream recipes. You will also need to heat all the ingredients either in the microwave or on the stovetop. Other things you will need are a glass or Pyrex bowl, measuring spoons, a mixing spoon and containers. Again, you will have most of these items in your kitchen.

Moisture Cream Ingredients

The ingredients used to make moisture creams include wax, oil, water, emulsifiers and preservatives. Moisture creams are emulsions made with wax, oil and water. It is easy to find beeswax, oils and water, however emulsifiers and preservatives are specialist ingredients, which are more difficult to source. Emulsifiers are used to create a stable emulsion between the oil phase and water phase. Without an emulsifier, you could not make an oil and water emulsion that would remain stable for any length of time. Preservatives are important in cream making to inhibit growth of bacteria, fungus and mould. Without a preservative, moisture cream would become contaminated with growth, making the cream unsafe to use. While emulsifiers and preservatives are quite specialist, today there are a number of online soap and toiletry supply companies which cater to hobbyists. You can buy these ingredients in small quantities for home use.

Try adding honey and other anti-oxidants to your moisture cream to make an age defying beauty cream. Buckwheat, ivy, heather and other dark honey varieties add polyphenols and emollients. You could also add vitamin E as another source of anti-oxidants. The advantage to making your own moisture cream is you can formulate a product to your needs and taste. You can personalise the cream by adding your favourite fragrance to the formula.

Making Moisture Cream

I was interested in making my own moisturising cream for quite a while, but I have to admit I was reluctant. In order to make creams, you need to make an emulsion with water and oil and if the emulsion is made incorrectly, the cream will separate. I am sure many of you have made oil and vinegar salad dressing. You may initially get the oil and vinegar to emulsify, but they easily separate, with the oil floating on top of the vinegar. Making an emulsion that will not separate, is the challenge when making your own moisture creams. Emulsifiers are used in cream making to facilitate the emulsion. Choosing a good emulsifier can mean the difference between success and failure.

Years ago the natural chemical, sodium tetraborate decahydrate, commonly known as borax, was used to emulsify moisture creams. Recipes would call for a pinch of borax added to the water. The borax containing water phase, could then be easily mixed into the oil phase of the cream recipe. Borax can be difficult to work with because borax powder is very fine and can easily become air borne. It can cause respiratory irritation if inhaled and eye irritation if the dust makes contact with eyes. Additionally, making emulsions with borax is not fool proof. You need to be very careful when mixing the oil and water or the two phases will separate. Because of this difficulty and the safety issues associated with using borax, borax is rarely used. When I began investigating emulsifiers, I found there to be so many that I became overwhelmed. Additionally, emulsifiers were very specialist ingredients that were not easy to find and if you did find a supplier, you had to buy a lifetime supply! This is no longer the case

Emulsifiers

Today there are innumerable emulsifiers available from cosmetic supply companies. You can buy emulsifiers in small quantities for use at home. Some are powders, some are liquids and some come in the form of emulsifying wax. Many of today's emulsifiers are blends of more than one chemical, which increases the likelihood of success in forming an emulsion. I have tested a number of emulsifying products, very often in wax form. While many are very good emulsifiers, I find an emulsifying wax which is a blend of ceteareth-20, cetearyl alcohol, glyceryl stearate and PEG-40 stearate to work very well. This blend contains all vegetable-derived ingredients, creates a good emulsion with a silky consistency and is incredibly easy to use. It is the ease of use that attracted me.

With many emulsifiers, you need to heat the oil and water phase separately to the same temperature, then mix the oil and water very precisely. I put all the ingredients in a bowl, melt and mix until the emulsion has formed. There is no need to measure temperatures or work with the two phases separately (oil phase and water phase). The recipes included in this book are formulated to use this simplified method with the emulsifying wax blend of ceteareth-20, cetearyl alcohol, glyceryl stearate and PEG-40 stearate. If you use a different emulsifying system, adjust the method to that of those suggested by the emulsifier's manufacturer. For instance, if the manufacturer recommends you heat the oil and water phase to a specific temperature, follow these instructions for optimal results. After the emulsion is made, additional ingredients can be added including; honey, fragrance, and preservative.

Preservatives

The final ingredient to add to your moisture cream is a preservative. Because of the high water content in cream emulsions, they are ideal growing places for moulds, fungi and bacteria. Adding a preservative serves to inhibit the growth of the unwanted organisms. Oil alone does not facilitate the growth of bacteria, mould or fungus. Water based products, however do support contamination and growth. If you add honey to your moisture creams, this will also facilitate the growth of contaminating microbes. I use a blend of phenoxyethanol and ethylhexylglycerin at a final concentration of 1%. There are other preservatives, which work just as well in moisture creams. If you use a different preservative, use the concentration recommended by the manufacturer. You can buy a variety of preservatives in small quantities from online soap and cosmetic supply companies.

Buckwheat Honey and Molasses Anti-oxidant Cream

A basic cream emulsion made with beeswax, sweet almond oil and cocoa butter and water, makes the most of the anti-oxidant activity of buckwheat honey and molasses.

Ingredients	Imperial	Metric
Unrefined Beeswax	0.3 oz	8 g
Cocoa Butter	0.8 oz	24 g
Sweet Almond Oil	4.5 oz	128 g
Emulsifying Wax	1.5 oz	42 g
Water	14.1 oz	400 g
Buckwheat Honey	½ tsp	2.5 ml
Molasses	3 ½ tsp	2.5 ml
Preservative	1 tsp	6 g

Making the basic honey and beeswax cream

1. Weigh beeswax, cocoa butter, sweet almond oil, emulsifying wax and water as you add them to a large glass bowl.
2. The glass bowl now contains all the ingredients except the buckwheat honey, molasses and preservative.
3. Transfer the glass bowl to the top of the bain-marie or put into the microwave.
4. Heat until the beeswax, cocoa butter and emulsifying wax have melted completely.
5. Ensure all the beeswax and other solids have completely melted.
6. Take care when you begin to stir the hot mixture. If the mixture is very hot, the water can boil when you begin to mix. It is best to use a bowl with some extra depth to contain the liquid if it should boil.
7. Remove from heat and stir the melted ingredients thoroughly with whisk or spoon.
8. Use the stick mixer or hand blender to mix the cream ingredients into an emulsion. As the cream cools, it will thicken. You do not need to stir the cream continuously, but should stir intermittently as the cream cools.
9. When the cream has cooled to the point where the mixture is beginning to look like whipped cream, add buckwheat honey, molasses and preservative.

You can add few drops of fragrance or essential oil if you would like to add some fragrance to the cream. If you do not have buckwheat honey, try using ivy or heather honey in your anti-oxidant moisture cream.

Witch Hazel and Propolis Anti-oxidant Cream

Ingredients	Imperial	Metric
Unrefined Beeswax	0.3 oz	8 g
Cocoa Butter	0.8 oz	24 g
Rape Seed Oil	3.5 oz	100 g
Olive Oil	1 oz	28 g
Emulsifying Wax	1.5 oz	42 g
Water	14.1 oz	400 g
Propolis	A few drops	
Witch Hazel	1 tsp	5 ml
Heather Honey	½ tsp	2.5 ml
Preservative	1 tsp	6 g

Preparation
1. Weigh beeswax, cocoa butter, rape seed oil, olive oil, emulsifying wax and water as you add them to a large glass bowl.
2. Transfer the glass bowl to the top of the bain-marie or pan of water or put into the microwave.
3. Heat until the beeswax, cocoa butter and emulsifying wax have melted completely.
4. Take care when you begin to stir the hot mixture. If the mixture is very hot, the water can boil when you begin to mix. It is best to use a bowl with some extra depth to contain the liquid if it should bubble.
5. Use the stick mixer or hand blender to mix the cream ingredients into an emulsion. As the cream cools, it will thicken. You do not need to stir the cream continuously, but should stir intermittently as the cream cools.
6. Remove from heat and stir the melted ingredients thoroughly with whisk or spoon.
7. When the cream has cooled to the point where the mixture is beginning to look like whipped cream, add witch hazel, propolis, heather honey and preservative.

Witch hazel and propolis cream helps control acne and oily skin. Moisturisers frequently increase oiliness and break outs on oily skin. Propolis will decrease the amount of sebum produced. The anti-bacterial properties in propolis can also help prevent acne, while witch hazel can help reduce redness and swelling. Propolis and heather honey add anti-oxidants to this recipe.

Appendix 1

Oven Temperatures & Units of Measure

Oven temperatures

Some very old recipes refer to oven temperature as "moderate" or "hot" in the cooking instructions. Use the table below as a guide to help with oven temperatures. You can use the table to convert between Celsius and Fahrenheit temperature scales and determining the setting for a gas mark oven.

Gas Mark oven temperatures are less precise descriptions of temperature. Using an oven, which measures temperature in degrees, either Fahrenheit or Celsius, gives you more precise indication than gas marks. If making an old-fashioned recipe that calls for a "cool" oven, it is difficult to know the exact temperature required. The recipe may work best anywhere between 275°F and 300°F. Gas Mark is also a bit ambiguous. However, people bake best with the oven they are most familiar with so I have listed Fahrenheit and Celsius in the recipes included in this book.

Oven Settings

There are also many types of ovens available, from the very simple to the very sophisticated. My oven has a number of combination settings using fan and a selection of heat sources; bottom, top or bottom and top heat. I have found that baking with honey I have the most success setting my oven to heat from the bottom and with the fan.

Temperature Conversion between Celsius and Fahrenheit

$C = (F - 32) / 1.8 \quad F = (C \times 1.8) + 32$

Table 7 Oven Temperatures

Oven	Fahrenheit F°	Celsius C°	Gas Mark
Very cool	225	110	
	250	130	
Cool	275	140	1
	300	150	2
Moderate	325	170	3
	350	180	4

Moderately hot	375	190	5
Hot	400	200	6
	425	220	7
Very hot	450	230	8
	475	240	9

Units of Measurement

Another consideration when cooking are the units used in recipes. I know personally if I find a recipe that looks good and it is written in units I am unfamiliar with I am often put of from trying the recipe. I have used the American System, the Imperial System and the Metric System in the recipes in this book. I hope this will simplify making the recipes. Below is a conversion chart that may prove useful to convert between the three systems.

Volumes and Capacities

The basic units of volume or capacity used in cooking are cups and spoons in the American System, fluid ounces in the Imperial System, and milliliters in the Metric System. The table below shows the conversion between the systems for measurements of volume or capacity. Additional tables are provided which provide the conversions for weight is also given.

Table 8

Cooking Volumes and Capacities		
Cups & Spoons	Fluid Ounces	Milliliters
I C	8 fl oz	240 ml
¾ C	6 fl oz	180 ml
½ C	4 fl oz	120 ml
¼ C	2 fl oz	60 ml
²/₃ C	5.3 fl oz	160 ml
¹/₃ C	2.7 fl oz	80 ml
I T	.5 fl oz	15 ml
I tsp	.I fl oz	5 ml
½ tsp	.05 fl oz	2.5 ml
¼ tsp	.025 fl oz	1.25 ml

Cooking weights

While the American System of cooking measurements using only volume or capacity, both the Imperial System and the Metric System of cooking use a combination or volumes and weights. In the Imperial System, for liquids, fluid ounzes are used to measure the amounts in volumes. For dry ingredients, ounces are used to measure the ingredient by weight. The Metric System also uses both capacity and weight, using milliliters for liquids and grams for solid ingredients. Below the table shows the Imperial and Metric Systems' units of weights and the conversion between ounces and pounds for Imperial and grams and kilograms for Metric.

Table 9

Cooking weights			
Imperial		Metric	
Ounces (oz)	Pounds (lb)	Grams (g)	Kilograms (kg)
I oz	$^1/_{16}$ lb	28.35 g	0.028 kg
16 oz	I lb	453.59 g	0.454 kg
0.035 oz	0.0022 lb	I g	0.001 kg
35.27 0z	2.2 lb	1000 g	I kg

The Imperial System for measuring weights uses ounces and pounds. There are 16 ounces in a pound. The Metric System uses grams and kilograms with 1000 grams in each kilogram. It is much more accurate to measure by weight than by volume because your measurements are more reproducible when made using a scale than using a container which holds volume. Because of this, cooking results are probably more consistent when the Imperial or Metric System's is used to measure the ingredients.

Table 10 Col

Water White	Extra White	White
Mint	Alsike Clover	Alfalfa
Mountain Laurel	Black Locust	Basswood
White Sweet Clover	Blue Thistle	Blue Curls
Yellow Sweet Clover	Crimson Clover	Blue Vine
	Hairy Vetch	Cape Vine
	Hubam Clover	Cotton
	Knapweed	Cucumber
	Lima Bean	Firewood
	Vetch	Gallberry
		Horsemint
		Manzanita
		Marigold
		Mesquite
		Purple Loosestrife
		Sage
		Sourwood
		Star Thistle

Table 10 lists a selection of honey samples and the colour of these honey varieties, as measured by White (1962). The honey colour can be used to estimate the levels of anti-oxidants in a honey variety. Even the lightest honeys (water white) have significantly higher levels of anti-oxidants than

y Varieties

ra Light Amber	Light Amber	Amber	Dark Amber
Aster	Athel Tree	Alfalfa-Honeydew	Buckwheat
Autumn Blend	Blueberry	Blackberry	Grape
Bergamot	Boneset	Cedar-Honeydew	Honeydew
Cantaloupe	Brown Knapweed	Chinquapin	
Eucalyptus	Cranberry	Coral Vine	
Goldenrod	Crotalaria	Hickory-Honeydew	
Heartsease	Goldenrod	Oak-Honeydew	
apanese Bamboo	Holly	Peppervine	
Palmetto	Mallow Weed	Raspberry	
Privet	Mexican Clover	Sumac	
Snowbrush	Mustard	Tulip Tree	
Spearmint	Pepperbush		
Spring Blossom	Peppermint		
ummer Blossom	Prune		
Tupelo	Sunflower		
Willow	Thyme		
Winter Blend	Titi		

sugar. As the honey colour darkens, the level of anti-oxidants in the honey increases. The varieties listed above are only a selection of the world's honey varieties. Use the colour guide on the back cover to help you choose dark honey varieties, with high anti-oxidants.

Bibliography

Alfonsus, E. C. (1983). Some sources of propolis. Glean. Bee Cult., 61: 92-93.

Aljadi, A. M. & Kamaruddin, M. Y.(2004). Evaluation of the phenolic contents and antioxidant capacities of two Malaysian floral honeys, Food Chemistry, 85: 513–518.

Al-Waili, N. S. (2003). Topical application of natural honey, beeswax and olive oil mixture for atopic dermatitis or psoriasis: partially controlled, single-blinded study. Complement Ther Med., 11(4):226-34.

Al-Waili, N. S. (2005). Clinical and mycological benefits of topical application of honey, olive oil and beeswax in diaper dermatitis. Clin Microbiol Infect., 11(2):160-3.

Ames, B. M., Shigena, M. K. & Hagen, T. M. (1993). Oxidants, antioxidants, and the degenerative diseases of ageing. Proceedings of the National Academy of Sciences USA, 90: 7915–7922.

Anghileri, L. J. & Thouvenot, P. (2000). Natural polyphenols– iron interaction. Biological Trace Element Research, 73: 251–258.

Baltrušaityte, V., Venskutonis, P. R. & Čeksterytė, V. (2007). Radical scavenging activity of different floral origin honey and beebread phenolic extracts. Food Chemistry, 101: 502–514.

Benzie, F. F. & Strain, J. J. (1996). The Ferric Reducing Ability of Plasma (FRAP) as a Measure of "Anti-oxidant Power": The FRAP Assay . Analytical Biochemistry, 239: 70–76.

Benzie, I. F. F. & Strain, J. J. (1999). Ferric reducing/antioxidant power assay: Direct measure of total antioxidant activity of biological fluids and modified version for simultaneous measurement of total antioxidant power and ascorbic acid concentration. Methods in Enzymology. 299: 15–27.

Beretta, G., Granata, P., Ferrero, M., Orioli, M. & Facino, R. (2005). Standardization of anti-oxidant properties of honey by a combination of spectrophotometric/fluorimetric assays and chemometrics. Analytica Chimica Acta, 533(2): 185-191.

Bertoncelj, J., Dobersek, U., Jamnik, M. & Golob, T. (2007). Evaluation of the phenolic content, anti-oxidant activity and colour of Slovenian honey. Food Chemistry, 105(2): 822-828.

Buratti, S., Benedetti, S. & Cosio, M.S. (2007). Evaluation of the anti-oxidant power of honey, propolis and royal jelly by amperometric flow injection analysis. Talanta, 71: 1387–1392.

Busserolles, J., Gueux, E., Rock, E., Mazur, A. & Rayssiguie Y. (2002). Substituting Honey for Refined Carbohydrates Protects Rats from Hypertriglyceridemic and Prooxidative Effects of Fructose. J. Nutr., 132(11):3379-82.

Campbell, M. G. (1913). A Textbook of Domestic Science for High Schools. Bedford, Massachusetts: Applewood Books.

Carlsen, M. H., Halvorsen, B. L., Holte, K., Bøhn, S. K., Dragland, S., Sampson, L., Willey, C., Senoo, H., Umezono, Y., Sanada, C., Barikmo, I., Berhe, N., Willett, W. C., Phillips, K. M., Jacobs, D. R. & Blomhoff, R. (2010). The total anti-oxidant content of more than 3100 foods, beverages, spices, herbs and supplements used worldwide. Nutrition Journal, 9:3.

Carpes, S. T., Mourão, G. B., De Alencar, S. M. & Masson, M. I. (2009). Chemical composition and free radical scavenging activity of Apis mellifera bee pollen from Southern Brazil Braz. J. Food Technol., 12(3): 220-229.

Chang, C. C., Yang, M. H., Wen, H. M. & Chern, J. C. (2002). Estimation of Total Flavonoid Content in Propolis by Two Complementary Colorimetric Methods. Journal of Food and Drug Analysis, Vol. 10(3):178-182

Chang, X., Wang, J., Yang, S., Chena, S. & Song Y. (2011). Antioxidative, antibrowning and anti-bacterial activities of sixteen floral honeys. Food Funct., 2: 541.

Chase, A. W. (1875). Dr Chase's Second Receipt Book. Toledo, Ohio: Chase Publishing Company.

Cooper, R, A. & Molan, P. C. (1999). Honey in wound care. Journal of Wound Care, 8: 340-342.

D'Arcy, B. R. (2005). Anti-oxidants in Australian Floral Honeys–Identification of health-enhancing nutrient components. A report for the Rural Industries Research and Development Corporation. Publication No 05/040.

Dattatraya, G. N., Vaidya, H. S. & Behera, B. C. (2009). Anti-oxidant properties of Indian propolis. Journal of ApiProduct and ApiMedical Science, 1(4): 110 – 120.

Dimins, F., Kuka, P. & Augspole, L. (2010). Characterisation of Honey Antioxidative Properties. International Conference of Food Inovation.

Dick, W. B. Encyclopedia of practical receipts and processes. (1872). New York: Dick & Fitzgerald Publishers.

Dragar, C. Antioxidant properties of Australian medicinal honeys, Journal of Agricultural and Food Chemistry, manuscript in preparation.

Farnesi, A. P., Aquino-Ferreira, R., De Jong, D. & Soares, A. E. E. (2009). Effects of stingless bee and honey bee propolis on four species of bacteria. Genet. Mol. Res., 8 (2): 635-640.

Ferrali, M., Signorini, C., Caciotti, B., Sugherini, L., Ciccoli, L., Giachetti, D. & Comporti, M. (1997). Protection against oxidative damage of erythrocyte membrane by the flavonoid quercetin and its relation to iron chelating activity. FEBS Letters, 416: 123–129.

Ferreira, I.C.F.R., Aires, E.; Barreira, J.C.M.; Estevinho, L.M. (2009). Antioxidant activity of Portuguesehoney samples: Different contributions of the entire honey and phenolic extract. Food Chem., 114: 1438-1443.

Frankel, S., Robinson, G. E. & Berenbaum, M. R. (1998). Anti-oxidant capacity and correlated characteristics of 14 unifloral honeys. Journal of Apicultural Research, 37: 27-31.

Gheldof N. & Engeseth N. J. (2002). Anti-oxidant Capacity of Honeys from Various Floral Sources Based on the Determination of Oxygen Radical Absorbance Capacity and Inhibition of in Vitro Lipoprotein Oxidation in Human Serum Samples. J. Agric. Food Chem., 50: 3050-3055.

Gheldof, N., Wang, X. H., & Engeseth,, N. J. (2002). Identification and Quantification of Anti-oxidant Components of Honeys from Various Floral Sources. J. Agric. Food Chem., 50 (21): 5870–5877.

Gheldof, N., Wang, X. H., & Engeseth,, N. J. (2003). Buckwheat honey increases serum antioxidant capacity in humans. J. Agric. Food Chem., 51: 1500-1505.

González, M., Guzmán, B., Rudyk, R., Romano, E. & Molina, M. A. A. (2003).

Spectrophotometric Determinationof Phenolic Compounds in Propolis. Lat. Am. J. Pharm., 22(3): 243-8.

Hamdy, A. A., Ismail, H. M., El-Moneim, A., AL- Ahwal, A. & Gomaa, N. F. (2009). Determination of Flavonoid and Phenolic Acid. Contents of Clover, Cotton and Citrus Floral Honeys. Egypt Public Health Assoc, 84(3 & 4).

Harman, D. (1956). Aging: a theory based on free radical and radiation chemistry. J Gerontol, 11(3):298-300.

Hegazi, A. G. & Abd El-Hady, F. K. (2009). Influence of Honey on the Suppression of Human Low Density Lipoprotein (LDL. Peroxidation (In vitro) eCAM, 6(1).113–121

Henriques, A., Jackson, S., Cooper, R. & Burton, N. (2006). Free radical production and quenching in honeys with wound healing potential. J. Antimicrob. Chemother. 58 (4): 773-777.

Inoue, K., Murayama, S., Seshimo, F., Takeba, K,. Yoshimura, Y. & Nakazawa, H. (2005). Identification of phenolic compound in manuka honey as specific superoxide anion radical scavenger using electron spin resonance (ESR) and liquid chromatography with coulometric array detection. J Sci Food Agric, 85:872–878

Inoue, S., Koya-Miyata, S., Ushio, S., Iwaki, K., Ikeda, M. & Kurimoto, M. (2003). Royal Jelly prolongs the life span of C3H/HeJ mice: correlation with reduced DNA damage. Exp Gerontol, 38:965-969.

Irish, J., Blair, S. & Carter, D.A. (2011). The Anti-bacterial Activity of Honey Australian Flora. PLoS ONE, 6(3): March.

Johnson, J. A., Miller, D. & White, J. W. Jr. Honey in your Baking Kansas State University Extension Service Circular, Published: 1959

Johnson, R. K., Appel, L. J., Brands, M, Howard, B. V., Lefevre, M., Lustig, R. H., Sacks, F., Steffen, L. M. & Wylie-Rosett,J. (2009). Dietary Sugars Intake and Cardiovascular Health. A Scientific Statement From the American Heart Association on behalf of the American Heart Association Nutrition Committee of the Council on Nutrition, Physical Activity, and Metabolism and the Council on Epidemiology and Prevention. Circulation, 120: 1011-1020

Kaškoniene, V., Maruška, A., Kornyšova, O., Charczun, N., Ligor, M. & Buszewski, B. (2009). Quantitative and qualitative determination of phenolic compounds in honey. Chemine Technologija, 3: 52.

Kawabata, T., Schepkin, V., Haramaki, N., Phadke, R. S., & Packer, L. (1996). Iron coordination by catechol derivative antioxidants. Biochemical Pharmacology, 51, 1569–1577.

Khalil, M. I., Mahaneem, M., Jamalullail, S. M. S., Alam, N. & Sulaiman, S. A. (2011). Evaluation of Radical Scavenging Activity and Colour Intensity of Nine Malaysian Honeys of Different Origin. Journal of ApiProduct and ApiMedical Science, 3(1): 04–11.

König, B. (1985). Plant sources of propolis. Bee World, 66(4): 136–139.

Krell, R. (1996). Value-Added Products Beekeeping. Agricultural Services Bulletin No. 124. Food and Agriculture Organization of the United Nations Rome.

Krpan M., MarKoVlc K., Šarlc G., HruŠKar M., & VaHclc N. (2009). Anti-oxidant Activities and Total Phenolics of Acacia Honey. Czech J. Food Sci., 27

Kuwabara, Y., Hori, Y., Yoneda, T. & Ikeda, Y. (1996). The antioxidant properties of royal jelly. Jpn Pharmacol Ther, 24:63-67.

LaChmaN, J., heJtmáNkoVá, A., Sýkora, J., karbaN, J., orSák, M. & rygeroVá, B. (2010). Contents of Major Phenolic and Flavonoid Anti-oxidants in Selected Czech Honey. Czech J. Food Sci., 28 (5): 412–426

Liu, F. L., Fu, W. J., Yang, D. R., Peng, Y. Q., Zhang, X. W. & He, J. (2004). Reinforcement of bee–plant interaction by phenolics in food. Journal of Apicultural Research, 43(4): 155–157.

Mademtzoglou, D., Haza, A.I,. Coto, A.L. & Morales, P. (2010). Rosemary, Heather And Heterofloral Honeys Protect Towards Cytotoxicity of Acrylamide In Human Hepatoma Cells. Revista Complutense de Ciencias Veterinarias, 4(2): 12-32.

Mandel, S. A., Amit, T., Zheng, H., Weinreb, O. & Youdim, M. B. H. (2006). The essentiality of iron chelation in neuroprotection: A potential role of green tea catechins. Oxidative Stress and Disease, 22: 277–299.

Mandel, S. A., Avramovich-Tiorsh, Y., Reznichenko, L., Zheng, H., Weinreb, O., Amit, T. & Youdim, M.B. (2005). Multifunctional activities of green tea catechins in neuroprotection. Modulation of cell survival genes, iron-dependent oxidative stress and PKC signaling pathway. Neurosignals, 14(1-2):46-60.

Manning, R. (2012). Research into Western Australian Honeys. Department of Agriculture and Food, Western Australia.

Marcucci, M.C., Ferreres, F., Garcia-Viguera, C., Bankova, V.S., De Castro, S.L., Dantas, A.P., Valente, P.H. & Paulino, N. (2001). Phenolic compounds from Brazilian propolis with pharmacological activities. J Ethnopharmacol, 74:105-112.

Miller, D., White, J. W. Jr. & Johnson, J. A. (1960). Honey Improves Baked Products Agricultural Experiment Station. Bulletin 411.

Mohamed, M., Sirajudeen, K. N. S., Swamy, M. Yaacob, N. S. & Sulaiman, S. A. (2010). Studies on the antioxidant properties of Tualang honey of Malaysia. African Journal of Traditional, Complementary and Alternative medicines, 7(1): 59-63.

Nakajima, Y., Tsuruma, K., Shimazawa, M., Mishima, S. & Hara, H. (2009). Comparison of bee products based on assays of anti-oxidant capacities. BMC Complementary and Alternative Medicine, 9:4.

National Honey Board & American Institute of Baking. (1990). Honey: Its Utilization in Bakery Products.

Özkök, A., D'arcy, B., & Sorkun, K. (2010). Total Phenolic Acid and Total Flavonoid Content of Turkish Pine Honeydew Honey. Journal of ApiProduct and ApiMedical Science 2 (2): 65 – 71.

Phillips, K. M., Carlsen, M. H. & Blomhoff, R. (2009). Total anti-oxidant content of alternatives to refined sugar. J Am Diet Assoc., 109(1):64-71.

Piljac-Žegarac, J., Stipcevic, T. & Belšcak, A. (2009). Antioxidant properties and phenolic content of different floral origin honeys. Journal of ApiProduct and ApiMedical Science, 1(2): 43 – 50.

Reddan, J. R., Giblin, F. J., Sevilla, M., Padgaonkar, V., Dziedzic, D. C., Leverenz, V. R., Misra, I. C., Chang, J. S. & Pena, J. T.(2003). Propyl gallate is a superoxide dismutase mimic and protects cultured lens epithelial cells from H2O2 insult. Experimental Eye Research,

76: 49–59.

Robb, S. J. (2009). Dr Sara's Honey Potions. Northern Bee Books.

Robb, S. J. & Connor, J. R. Nitric oxide protects astrocytes from oxidative stress. (2002). Ann. N.Y Acad. Sci., 962: 93-102.

Robb, S. J., Gaspers, L. D., Wright, K. J., Thomas, A. P. & Connor, J. R. (1999). Influence of nitric oxide on cellular and mitochondrial integrity in oxidatively stressed astrocytes. J Neurosci Res., 56(2):166-76.

Robb S. J., Robb-Gaspers, L. D., Scaduto, R. C. Jr., Thomas, A. P. & Connor, J. R. (1999). Influence of calcium and iron on cell death and mitochondrial function in oxidatively stressed astrocytes. J Neurosci Res., 55(6):674-86.

Robb, S. J. & Connor, J. R. (1998). An in vitro model for analysis of oxidative death in primary mouse astrocytes. Brain Res., 788(1-2):125-32.

Sahinler, N., Gül, A. & Sahin, A. (2005). Vitamin E supplement in honey bee colonies to increase cell acceptance rate and royal jelly production. Journal of Apicultural Research 44(2): 58–60.

Salonen, A., Hiltunen, J. & Julkunen-Tiitto, R. (2011). Composition of Unique Unifloral Honeys from the Boreal Coniferous Forest Zone: Fireweed and Raspberry Honey. Journal of ApiProduct and ApiMedical Science 3(3): 128 – 136.

Seijo, M. C., Jato, M. V., Aira, M. J. & Iglesias, I. (1997). Unifloral honeys of Galicia. Journal of Apicultural Research, 36(3/4): 133–139.

Sestili, P., Guidarelli, A., Dacha, M. & Cantoni, O. (1998). Quercetin prevents DNA single strand breakage and cytotoxicity caused by tert-butylhydroperoxide: Free radical scavenging versus iron chelating mechanism. Free Radical Biology and Medicine, 25: 196–200.

Sestili, P., Diamantini, G., Bedini, A., Cerioni, L., Tommasini, I., Tarzia, G. & Cantoni, O. (2002). Plant-derived phenolic compounds prevent the DNA single-strand breakage and cytotoxicity by tert-butylhydroperoxide via an iron-chelating mechanism. The Biochemical Journal, 364: 121–128.

Sharpe, M. A., Robb, S. J. & Clark, J.B. (2003). Nitric oxide and Fenton/Haber-

Weiss chemistry: nitric oxide is a potent anti-oxidant at physiological concentrations. J Neurochem., 87(2):386-94.

Srisayam, M. & Chantawannakul, P. (2010). Antimicrobial and anti-oxidant properties of honeys produced by Apis mellifera in Thailand. Journal of ApiProduct and ApiMedical Science, 2(2): 77 – 83.

Tan, H. T., Rahman, R. A., Gan, S. H., Halim, A. S., Hassan, S. A., Sulaiman, S. A. & Kirnpal-Kaur, B. S. (2009). The antibacterial properties of Malaysian tualang honey against wound and enteric microorganisms in comparison to manuka honey. BMC Complementary and Alternative Medicine, 9: 34.

Tamura, T., Fujii, A. & Kuboyama, N. (1987). Antitumor effects of royal jelly. Nippon Yakurigaku Zasshi, 89:73-80.

Taormina, P. J., Niemira, B. A. & Beuchat, L. R. (2001). Inhibitory Activity of Honey Against Foodborne Pathogens as Influenced by the Presence of Hydrogen Peroxide and Level of Anti-oxidant Power. International Journal of Food Microbiology 69: 217-225

Tosi, E. A., Ciappini, M. C., Cazzolli, A. F., & Tapiz, L M. (2006). Physico Chemical Characteristics Of Propolis Collected In Santa Fe (Argentine). Apiacta, 41: 110-120.

United States Department of Agriculture Database for the Oxygen Radical Absorbance Capacity (ORAC) of Selected Foods, Release 2. 2010. Prepared by Nutrient Data Laboratory, Beltsville Human Nutrition Research Center (BHNRC), Agricultural Research Service (ARS), U.S. Department of Agriculture (USDA).

United States Department of Agriculture. (2010). National Nutrient Database for Standard Reference, Release 23.

United States Department of Agriculture and United States Department of Health and Human Services. (2010). Dietary Guidelines for Americans, 7th Edition, Washington, DC: U.S. Government Printing Office.

United States Deepartment of Agriculture. (2010). The Food Guide Pyramid. A Guide to Daily Food Choices. Home and Garden Bulletin Number 252.

United States Department of Agriculture. Agricultural Research Service. (2011). USDA National Nutrient Database for Standard Reference, Release 24.

United States Department of Agriculture. (2007). Oxygen Radical Absorbance Capacity (ORAC) of Selected Foods – 2007.

United States Department of Agriculture & Arkansas & Children's Nutrition Center. (2007). Oxygen Radical Absorbance Capacity (ORAC) of Selected Foods. Nutrient Data Laboratory, Beltsville Human Nutrition Research Center, Agricultural Research Service.

Venugopa,I S. & Devarajan, S. (2011). Estimation of total flavonoids, phenols and anti-oxidant activity of local and New Zealand manuka honey. Journal of Pharmacy Research, 4(2).,464-466.

van den Berg, A. J. J., van den Worm, E., Quarles van Ufford, H. C., Halkes, S. B. A., Hoekstra, M. J. & Beukelman, C. J. (2008). An in vitro examination of the anti-oxidant and anti-infl ammatory properties of buckwheat honey. Journal of Wound Care, 17(4)..

Wilczyńska A. (2010). Phenolic Content And Anti-Oxidant Activity Of Different Types Of Polish Honey – A Short Report. Pol. J. Food Nutr. Sci., 60(4): 309-313.

Wang, X. H., Gheldof, N. & Engeseth, N. J. (2004)., Effect of Processing and Storage on Anti-oxidant Capacity of Honey. Journal of Food Science, 69: 96–101.

White, J. W. Jr. & Doner L. W. (1980). Honey Composition and Properties. In: Beekeeping in the United States. Agricultural Handbook 335. Science and Education Administration. Superintendent of Documents, US Government Printing Office, Washington, D.C.

White, J. W. Jr. & Subers, M. H. (1963). Studies on Inhibine. 2. A Chemical Assay. Journal of Apicultural Research, 2(2). :93~100.

White, J. W. Jr. Riethof, M. L., Subers, M. H. & Kushnir, I. (1962). Composition of American Honeys. Technical Bulletin No. 1261. Agricultural Research Service. Superintendent of Documents, US Government Printing Office, Washington, D.C

White, J. W. Jr. (1962). Composition of American Honeys Agricultural Research Service.

White, J. W. Jr. (1961). A Survey of American Honeys. 5. Effect of Area of Production on Composition. Gleanings in Bee Culture, March.

White, J. W. Jr. (1967). Composition of Honey. VII. Proteins. Source: Journal of Apicultural Research, Vol. 6(3): 163-178.

White, J. W. Jr. Subers, M. H. & Schepartz, A. I. (1963). The identification of inhibine, the antibacterial factor in honey, as hydrogen peroxide and its origin in a honey glucose-oxidase system. Biochimica et Biophysica Acta 73: 57-70.

Wilczyńska, A. (2010). Phenolic Content And Anti-Oxidant Activity Of Different Types of Polish Honey – A Short Report Pol. J. Food Nutr. Sci., 60(4): 309-313

Yoshino, M., & Murakami, K. (1998). Interaction of iron with polyphenolic compounds: Application to antioxidant characterization. Analytical Biochemistry, 257: 40–44.

Yaoa, L., Jianga, Y., Singanusong, R., Datta, N. & Raymont, K. (2005). Phenolic acids in Australian Melaleuca, Guioa, Lophostemon, Banksia and Helianthus honeys and their potential for floral authentication. Food Research International, 38(6): 651–658.

Index